Facilities Management and the Business of Managing Assets

The importance of facilities management to the effective operation of all businesses is now widely accepted. Where there continues to be debate is what constitutes a successful approach, and how much attention it should be given within an overall business plan.

Drawing on both research and current practice, this book provides a systematic, innovative and business-focused approach to the management of facilities assets. The reader will discover why and how to use facility assets to achieve business goals and strategies for aligning them as a resource.

Striking a balance between management and technical aspects, the book covers:

- the basics of facilities asset management and the key elements of a systematic management approach;
- the key supporting capabilities for facilities management as a business function;
- a framework for considering strategic alignment of facilities assets and associated services with business needs; and
- the role of life-cycle asset management and its contribution to business resource management.

The message of this book and the benefits it can bring to businesses everywhere make it essential reading for executives as well as facilities managers. Its detailed explanations of all the key concepts involved, and lists of recommended further reading, also make it an excellent resource for those new to the industry and for students of property or facilities management.

Danny Then Shiem-Shin is Joint Coordinator CIB Working Commission W070 for Facilities Management and Asset Maintenance. Danny's academic career spans over three universities (Heriot-Watt University, Queensland University of Technology and Hong Kong Polytechnic University) where he developed postgraduate programmes in facilities management in departments with a focus on construction/property management and building services engineering. He has published and presented more than 70 papers internationally and co-authored *Facilities Management and the Business of Space* (2001).

Tan Teng Hee is an electrical engineer with 37 years' experience in electrical engineering, maintenance and facilities management across a diverse range of facilities including a large cultural centre, military facilities, hospitals, university campus, and government and commercial office buildings. He worked for the public and private sectors in Malaysia and Australia. Currently he is a facilities management advisor and lectures as a visiting lecturer.

Facilities Management and the Business of Managing Assets

Danny Then Shiem-Shin and
Tan Teng Hee

Routledge
Taylor & Francis Group
LONDON AND NEW YORK

First published 2013
by Routledge
2 Park Square, Milton Park, Abingdon, Oxon, OX14 4RN

Simultaneously published in the USA and Canada
by Routledge
52 Vanderbilt Avenue, New York, NY 10017

First issued in paperback 2020

Routledge is an imprint of the Taylor & Francis Group, an informa business

British Library Cataloguing in Publication Data
A catalogue record for this book is available from the British Library

Library of Congress Cataloging in Publication Data
Then, Danny Shiem-Shin.
Facilities management and the business of managing assets / Danny Then Shiem Shin and Tan Teng Hee.
p. cm.
Includes bibliographical references and index.
1. Facility management. 2. Strategic planning. I. Hee, Tan Teng. II. Title.
HF5547.2.T44 2013
658.2--dc23
2012028048

ISBN 13: 978-0-367-57654-7 (pbk)
ISBN 13: 978-0-415-27494-4 (hbk)

Typeset in Galliard
by Taylor & Francis Books

Contents

List of illustrations

Figures

Tables

Foreword

Everything you ever wanted to know about 'facilities management' but were afraid to ask would have been an alternative title for this book. Facilities management (FM) is a complex subject and does not lend itself to sound bites or elevator pitches. The term is often misapplied to a whole range of activities – from janitorial services to equipment maintenance.

Prof. Danny Then and Tan Teng Hee successfully set out not only a comprehensive explanation of FM for the uninitiated corporate director, but also how a strategic approach to the management of physical assets has a very real impact on the bottom line of an enterprise.

Just as facilities management is often a euphemism for outsourcing, so asset management is often erroneously thought of as a separate discipline from facilities management. In fact, facilities management requires an integrated approach to managing assets, resources and information, the procedures for which need to be absolutely aligned if an organization desires ultimate efficiency in its business support services. Whether the perfect solution is arranged in-house, outsourced or more likely a mixture of both depends on the specific requirements of each organization and the nature of the assets to be managed.

The title of this book cannot fully convey its rich insights into the methodology and processes required in managing facilities – not only the built and physical assets, but also the context of how best to support the objectives of an enterprise, whether in the private or public sector. For anyone seeking to understand the value proposition of facilities management, the co-authors of this book explain in clear terms the complexity of the subject, supported by understandable diagrams and comprehensive references.

Although it is written with the focus on managing assets – infrastructure, buildings, plant and equipment, furniture, in fact anything that has a tangible and/or monetary value to the organization – it also covers the many interrelated processes and systems in the wider field of FM.

The authors represent a combination of a respected academic and an experienced practitioner, with a rare mix of research, practical and technical experience accumulated over many years. This book not only demonstrates their depth of understanding of the subject matter, but also imparts their knowledge in a practical and logical format for readers to follow and implement in their management role. Their experience is not restricted to a narrow view of the subject of asset management, but reaches international best practice in the field of facilities management and technical and technological advances, all of which is conveniently referenced here.

Advances in technology applications and business improvement processes, aligning practical issues with business objectives, asset utilization and life-cycle planning are all covered here, making the book an invaluable handbook for the manager, a guide for the executive and a textbook for the student.

Lionel Prodgers
Director
Agents4RM International

Preface

The focus of this book is drawn from the perspective of facilities as a business resource with some unique characteristics. As high-value and long-life assets with significant operating and other costs, they deserve effective management in order to optimize their contribution to organizational objectives as well as to manage their costs and risks. Effective management should occur not only at the operational facilities management level but also at more senior management levels. With this in mind, we have identified as our primary intended audience senior managers at corporate level who wish to develop a more detailed knowledge of facilities asset management without necessarily delving into the specific processes and methods involved in operational management. At the same time, practising facilities managers will also find this a useful reference on the scope and key elements of facilities asset management that can guide their development of operational processes, systems and methods to suit their needs.

The contents of the book have been deliberately kept at a level that covers the main points of the various aspects of facilities asset management, but in sufficient detail to enable understanding and stimulate further enquiry as desired. It is hoped that such understanding would encourage more interactive dialogue between senior managers and their facilities managers and lead to better facilities and business outcomes. While we believe we have covered the subject matter in sufficient depth and scope, the book makes no claim to be a comprehensive resource on facilities management. Moreover, it was not intended to be one.

There is growing recognition of the need to improve the integration of facilities management within the broader context of managing organizational resources towards the achievement of strategic objectives. We have therefore endeavoured as much as possible to draw the links between facilities and business in an effort to emphasize the impact and contribution of facilities to business outcomes and the need for a structured and effective framework to manage them.

Much of the material that sets the scene for our discussions in the book has been drawn from the works of others, referenced accordingly. The rest of the material, which forms the bulk of the book, is drawn from our collective knowledge and experiences in the academic sphere on one hand and practical experience in the field of facilities and asset management on the other. Each has complemented the other, enhanced through interchange of ideas with other practitioners and perusal of related literature over many years in different countries. Any weaknesses or gaps in the concepts, frameworks and presentation are our responsibility alone.

Our acknowledgements and thanks go to the many friends, colleagues, clients, fellow practitioners, students and staff in industry (both public and private sectors) and academia with whom and from whom we have had the opportunity to discuss, debate and learn, over the 30-odd years that we have been operating in this field. A special thank you to Palm (Akarapong Katchamart) for his critiques and assistance during his short attachment at Hong Kong Polytechnic University. Our acknowledgements also go to our publishers who have been very patient and understanding over the period that it has taken for this book to take shape and substance.

Finally, but not least, our heartfelt thanks to our respective families who have been very supportive of our literary aspirations and tolerated with good humour the inevitable hours spent at the keyboard and buried in books and papers.

Danny Then Shiem-Shin and Tan Teng Hee
June 2012

1 Introduction

The creation of suitable workplace environments to support employees in performing their tasks is becoming a key issue for many organizations faced with constant change. Extraordinary change has already occurred, and will continue to impact on our lives and work, in such areas as technology, demography, work methods and the environment – both locally and globally. As far back as the 1980s and early 1990s, a number of research reports in the area of real estate and facilities management in North America, Europe and the United Kingdom highlighted a growing awareness of the critical role of corporate real estate assets and the need for more effective management (Zeckhauser and Silverman 1983, Veale 1989, Avis, Gibson and Watts 1989, Lloyd 1992, Arthur Anderson 1993, Jorroff *et al.* 1993, Then 1994, Apgar and Bell 1995, Gallup 1996). The global financial crisis in recent years has forced most organizations to optimize their scarce resources to meet business objectives. The need to align business resources to core business requirements has resulted in taking a more strategic view of their organizations' operational needs. This book is our attempt to provide a clear framework for senior corporate managers to consider facilities and assets issues as an integral component of overall strategic business planning.

1.1 Facilities and assets

The first challenge in writing a book on facilities management is establishing a clear structure and meaning for the terms that are frequently used throughout the book. This is, first, to achieve consistency and, second, to reduce any confusion that the reader may experience when different terms with similar meanings are used interchangeably. For example, 'assets' and 'facilities' are often used interchangeably. In this book we have chosen to use 'asset' as a general term referring to a business resource that has a financial value and is capable of providing some form of benefit (e.g. a financial return or a service) to the owner and users.

Figure 1.1 illustrates the structure that we have chosen to use as the basic building block of the terminology in this book.

In the context of this book, the term 'facility' refers to a group of buildings, the site or land on which they are located and the site improvements supporting the buildings. Consistent with this approach, we refer to places such as schools,

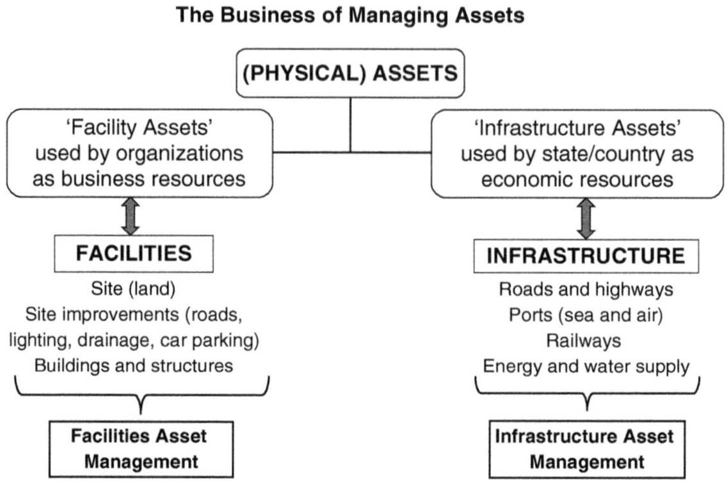

Figure 1.1 Facilities and infrastructure

military bases and hospitals, etc. as 'facilities'. An office building is referred to as an 'office facility' and a manufacturing plant is referred to as a 'manufacturing facility'. A portfolio consists of a group of facilities owned or operated by an organization (Figure 1.2). For example, a Department of Health will own or operate a portfolio of medical facilities. Roads, highways, ports (sea and air), railways and energy and water supply are referred to as 'economic infrastructure'. Both 'facilities' and 'economic infrastructure' are 'assets' or 'physical assets' as a generic group. By adopting the title *Facilities Management and the Business of Managing Assets*, we have intentionally drawn the boundaries of the book around facilities and the assets that make up the facilities. Undoubtedly many of the strategies and processes referred to in this book would apply equally well to economic infrastructure. However, in general, the management of economic infrastructure is outside the scope of this book.

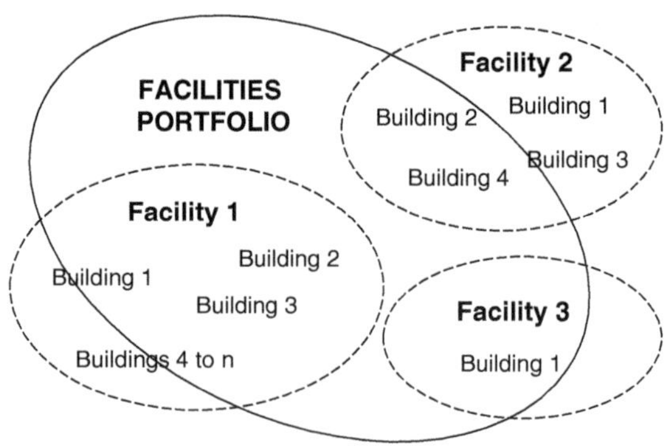

Figure 1.2 Facilities portfolio concept

1.2 A corporate perspective of facilities and assets

The business environments of both public and private sectors dictate the changes in the size and configuration of facilities portfolios based on the dynamics of demand and supply and the prevailing economic conditions which influence the supply of capital for investment and sustainability of operating costs. The public sector is a major owner and user of facilities as a platform for the delivery of services to the community. Health, defence and education sectors alone account for a significant portion of the total public sector portfolio. There has been an ongoing focus on public sector organizations optimizing their utilization of facilities and improving their management in pursuit of organizational cost efficiencies and in response to the challenges of constraints on capital funding opportunities to meet the growth in service demand. Similarly, the private sector is constantly under pressure to reduce costs and increase profitability and shareholder value. Facilities, as capital assets and operational cost-centres, will therefore also be subjected to closer scrutiny in terms of their capital value, operating costs and overall performance in contributing to the achievement of business objectives. Consequently there is a growing recognition of facilities as a business resource and the need for management from a business perspective rather than relegating their management to the domain of technical professionals operating at arm's length from the core business (NRC 2004, Apgar 1995, Then 1994).

The theme of this book is centred on the concept of buildings as facilities assets and their role as one of the key business resources employed to deliver business outcomes (the others being human, financial and information technology resources). Optimization of the contribution of facilities as a business resource through the alignment of facilities assets and their management with the needs and objectives of the organization is fundamental to effective resource management (Then 2003, CoreNet Global 2004, Nutt 2004, Then and Tan 2010).

Given that both the public and private sectors grapple with similar issues, albeit with different business objectives, the material in this book is aimed at being of benefit to facilities and business managers in both sectors. This book is therefore aimed at public and private sector corporate-level business managers and professionals in the construction, property and facilities management industries who wish to better understand the role of facilities and the key aspects of facilities asset management that have business implications for both public and private sectors.

1.3 Facilities management

Facilities management has an important business support role which contributes to aspects of business such as risk management, productivity, recruitment and retention, maximizing the value of capital investments in assets, managing the costs of facilities and addressing corporate responsibility issues such as environmental sustainability and climate change. Thus, it is in the best interests of an organization to embrace facilities management as an important business support function that provides facilities solutions to satisfy business needs (Varcoe 1992, Then 1994, Lindholm, Gibler and Leväinen 2006, CRC 2007, i-FM.net 2011).

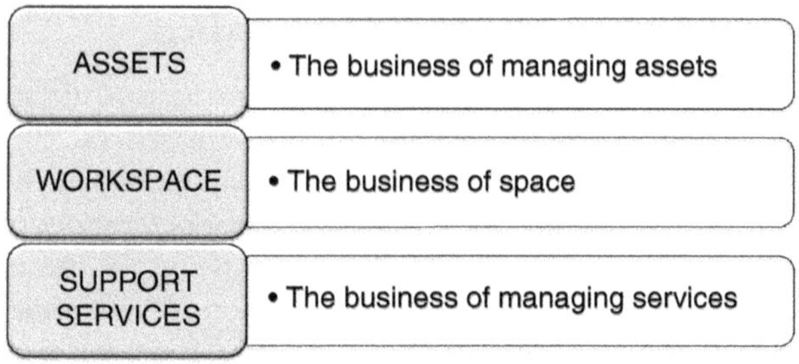

Figure 1.3 Management aspects in facilities management

There are many definitions of facilities management that have been proposed by professional institutions around the world (see Appendix 1). We have found them to be unhelpful from the business point of view, especially from the perspective of senior management, as the focus tended to be on required competences rather than what management should focus on. From our experience, the core of facilities management can be considered as management of three key related aspects of corporate facilities – assets, workspace and support services – all three of which are impacted by technological developments (Figure 1.3).

In relation to the three management aspects, the last two decades have seen major fundamental changes on the demand and supply sides of the facilities management market that have resulted in shifts in the management approach and focus, as shown in Figure 1.4.

The aspect of workplace management was the subject of another book by one of the authors, titled *Facilities Management and the Business of Space* (McGregor and Then, Arnold, 1999; Butterworth-Heinemann, 2001). This book aims to address issues and solutions relating to 'the business of managing assets' under the broad umbrella of 'facilities asset management'.

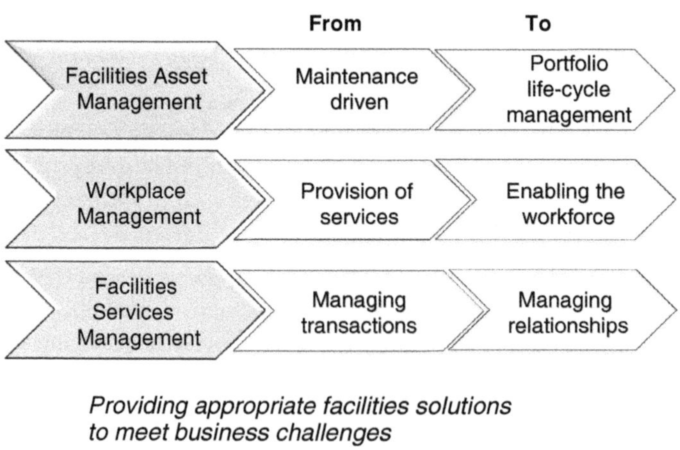

Figure 1.4 Changing management focus

Facility assets are viewed as business resources providing work and production environments and linked to specific business-related objectives. In order for these objectives to be fulfilled, both management and technical processes are required to be professionally applied within an appropriate policy and operational environment. These processes are well developed and generally referred to as physical asset management (e.g. NRC 1998, Hastings 2010, Lloyd 2010, Haynes and Nunnington 2010, Cotts and Rondeau 2003). Similarly, there is also ample literature on general aspects of facilities management processes (e.g. Finch 2012, Cotts, Roper and Payant 2010, Theriault 2010, Booty 2009, Barrett and Baldry 2003, Langston and Lauge-Kristensen 2002, Alexander, Ed. 1998) or specific aspects like workspace management (e.g. Best, de Valence and Langston, Eds 2003, McGregor and Then 2001, Becker and Steele 1994, Becker 1990) and facilities services management (e.g. Atkin and Brooks 2009, Alexander, Atkin, Brochner and Haugen, Eds 2004, Hinks and Reuvid 2002).

We have used the term 'facilities asset management' to emphasize the relationship and relevance of these processes to facilities and the assets that comprise the facility. Inherent in this approach is the adoption of an integrated and continuous life-cycle approach to asset management as opposed to a regime of *ad hoc* and reactive responses which can lead to ineffective capital investment, high operating costs, poor asset condition and a mismatch of facilities with actual needs.

Figure 1.5 illustrates the facility assets context across the facility life cycle, the relevant processes and the key resources required to manage them, within the overall umbrella of facilities management and underpinned by a drive for life cycle sustainability. Business drives the need for facility assets which evolve through a life cycle of planning, acquisition, utilization and maintenance and

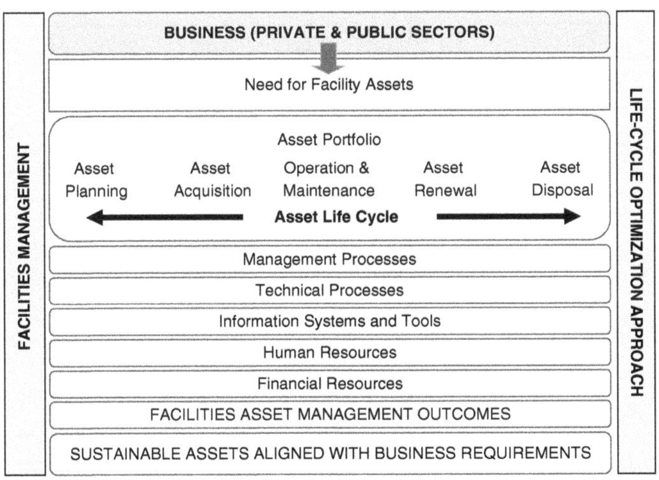

Figure 1.5 Facility assets context

finally to renewal or disposal. Management and technical processes are required to manage them at each stage of their life cycle and these processes need to be supported by accounting and facilities management systems as well as human and financial resources.

There is ample literature on physical asset management processes (e.g. NRC 1998; Hastings 2010; Lloyd, 2010) and it is not our intention to delve into it in depth. Our aim is to draw attention to some of the key processes and individual process elements that business and facilities managers need to focus on from a business perspective. Taking a 'business perspective' of these processes helps to identify those processes that have a strategic impact on business and hence focus corporate management attention on whether the physical outcomes of facilities asset management positively influence the alignment of facility assets with business needs and objectives. Figure 1.6 broadly illustrates the business perspective that complements the process-driven perspective.

1.4 Asset life cycle

Buildings in general are durable assets and last many years. The building life cycle has several stages and they need to be managed accordingly in order to derive the optimum value from the asset at the most economical cost over its life cycle (Langston 2005, Boussabaine and Kirkham 2004, Oschrin 2004). More importantly, in terms of asset management, each stage of the life cycle has implications for the following stages. Buildings are also not a homogeneous product in that they are normally made up of a large collection of components with varying

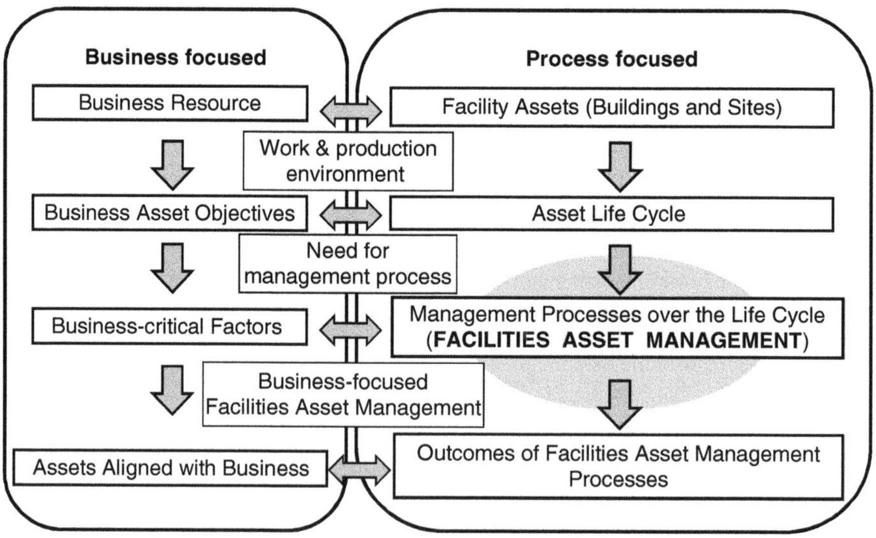

Figure 1.6 Business and process perspectives

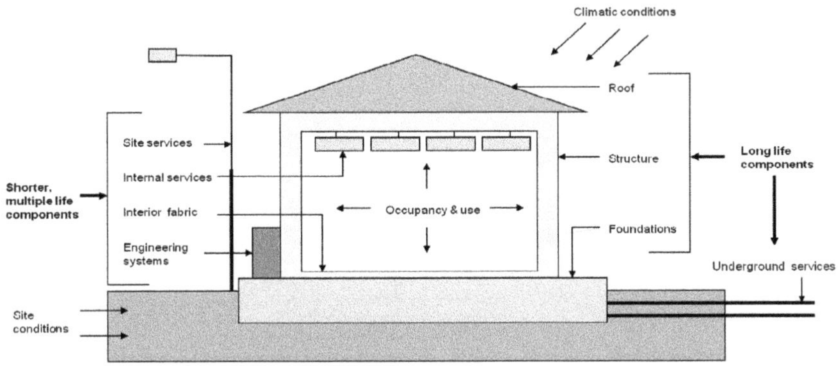

Figure 1.7 Layers of longevity of building components (Adapted from Brand 1994)

individual lifespans. Figure 1.7 illustrates the different components of a building asset and their varying longevities.

Each component's durability is influenced by its design and manufacture or construction, climatic exposure, the maintenance regime adopted and the manner in which it is used or even abused. Some components such as the exterior facade gradually deteriorate due to exposure to climatic conditions and other stresses (e. g. vandalism), while structural elements may fail gradually or suddenly (for example as a result of subterranean forces acting on the foundations). Internal fit-out and fabric deteriorates due to the wear and tear of occupancy. Engineering services on the other hand operate almost continuously and generally have far shorter lives than structures. A good analogy of the nature of building assets is given by Brand (1994) when he describes a building in terms of 'layers of longevity of building components'.

The significance of the life-cycle characteristics of building assets lies in the need for a whole-of-life approach to their management as a functioning business resource. While the planning and construction phases often occupy senior management attention, due to the significant capital investment and the euphoria of a new asset, very little of their attention is focused on the productive phase after construction when the economic benefits are realized over a period of 30 to 40 years or more. This is also the phase when the building assets will demand periodic injections of capital reinvestment in terms of component renewals, in addition to operating and maintenance costs (Figure 1.8).

It has been estimated that 85 to 90 per cent of a building's lifetime costs occur after it has been constructed (NRC 1998). In this respect, any attempts towards effective life-cycle management must take a holistic approach that introduces sustainable life-cycle costs strategies at the planning and design stage and strategies to optimize the value contribution of facilities throughout the post-construction productive phase up to disposal.

In the context of this book we have expressed the asset life cycle across five main stages. At each stage there are various technical and management processes applicable to manage the asset as it progresses from one stage to another.

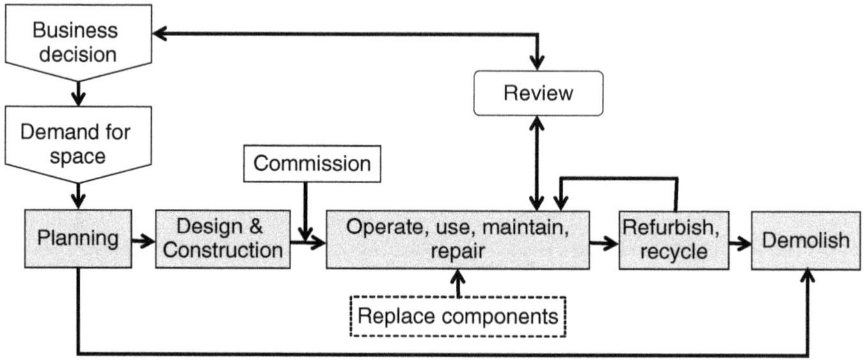

Figure 1.8 Asset life-cycle model

1.4.1 Planning stage

This is the stage of planning for one or more facility assets on a single site (a facility) or multiple sites (portfolio of facilities). It is intrinsically an exercise in determining the demand and supply situation and developing options for adjustment to the supply side of the equation to meet the demand for space as a consequence of anticipated business directions. If one of the outcomes is to acquire new assets, then planning for the acquisition of the assets commences and leads to the next stage.

1.4.2 Design and construction stage

This is the capital investment and asset creation stage (following the planning stage) with the objective of implementing the plan to develop, design and construct and transitioning the asset as an operational business resource. The result is a fully constructed asset ready to be occupied and operated and requiring maintenance from that point onwards. This stage determines the future operating costs of the asset to a significant extent and justifies the input of construction and facilities management experts to provide advice on construction and factors such as maintainability, operational efficiency and effectiveness and environmental sustainability issues.

1.4.3 Utilization and maintenance stage

This is the stage when the asset has transitioned from the acquisition stage to occupancy and use. It is the period when value (economic benefit) is returned to the organization from its capital investment in the asset. This is by far the longest phase of the asset's life and has the greatest physical, functional and financial impact on business processes and outcomes. The corporate guidelines or policies on space planning and allocation will determine the level of utilization of spaces with a facility asset. Utilization of the asset (amongst other factors like design) determines the level of maintenance required and this in turn has implications for the life of the asset, operational costs, risks and longer-term liabilities associated with asset condition and the need for renewal. Capital reinvestment is also required for replacements of components at the end of their useful lives.

1.4.4 Renewal stage

This is the stage where the facility asset no longer meets requirements (partially or fully) due to various factors (deteriorated condition, change in functional requirements, decline or increase in space demand) and decisions need to be made on options for end-of-life components, upgrades, refurbishment and/or expansion. The asset may also be reconfigured for a new use (e.g. conversion of old factory premises into residential apartments).

1.4.5 Disposal stage

This is the stage where the decision is made that the asset has reached the end of its useful life, renewal options are not physically and/or financially viable and heritage preservation considerations do not apply, and the asset is demolished. The products of demolition are disposed of as waste or recycled (e.g. timber).

Clearly the above stages are sequential in terms of a single facility asset. However, a group of facility assets within a facility or portfolio may all be at different stages of their life cycles at any one time, emphasizing the importance of and need for robust portfolio-wide, life-cycle asset planning to maintain the alignment of facilities with business directions.

1.5 Facilities asset management

An NRC publication (2004, p. 32) defined *facilities asset management* as:

> A systematic process for maintaining, upgrading, and operating physical assets cost effectively. It combines engineering principles with sound business practices and economic theory and provides tools to facilitate a more organized, logical approach to decision making.
>
> (Quoting from FHWA 1999, p. 7)

The importance of a facilities asset management approach is that it allows organizations to integrate facilities considerations into corporate decision-making and strategic-planning processes. This is a significant shift from past practice, whereby facilities-related decisions were often made after the organization's strategic direction had been set (NRC 2004, p. 32).

Figure 1.9 illustrates how a facilities asset management approach allows organizations to forge a direct link between organizational goals, facilities investment decisions and day-to-day operations.

The goal of facilities asset management is to give an organization the work environments it needs to achieve its missions by optimizing available resources. In

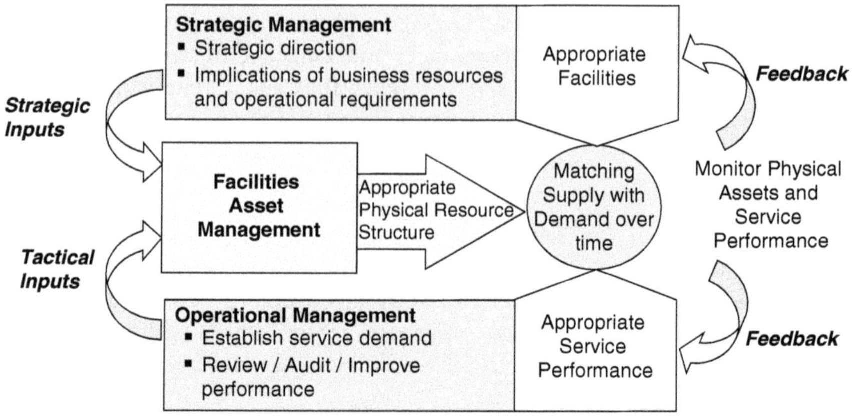

Figure 1.9 Linking organizational goals with facilities investment and operations (Adapted from Then 1996, 2003; NRC 2004, p. 33)

practice we see facilities asset management as a framework of management and technical processes implemented through the application of human and financial resources to manage facility assets from planning to disposal. The framework incorporates the principles of life-cycle management of physical assets in that decisions made in any stage have consequences for the following stages and therefore need to be carefully considered. In the context of this book we define facilities asset management in terms of four main process components (reflecting the asset life cycle), with each one having a range of individual process elements. The four main process components of facilities asset management are:

- strategic facilities review;
- strategic facilities planning;
- strategic facilities alignment;
- facilities utilization and management.

Within an organizational context, the processes are continuous and iterative, constantly responding to the driving corporate strategy which provides the inputs and controls for the management and technical processes to be applied. More detailed discussion of facilities asset management is provided in Chapter 3.

1.6 Terminology

In this book various terms are used repeatedly, as they are applied in the context of each section due to the fact that they are fundamental to the discussion as a common thread throughout the book.

We have used *organization* as a generic term for a corporate entity in the private sector as well as the public or government sector. That is, an entity that produces goods and services for commercial profit or as a result of tax-payer funding to serve the community. The term *business* is also used to represent such entities in as much as they conduct the 'business activities' necessary to produce the desired outputs and outcomes. *Business managers* and *executives* are used to refer, in a broad sense, to all those managers responsible for various aspects of the business, but excluding facilities management. They are in fact *internal clients* of the facilities management function. *Senior business managers* and *executives* refer to those corporate executives who have responsibility for the business at the Board of Directors level. A *facilities management department* refers to an internal business unit, section or division that performs facilities management functions for the organization. An external facilities management organization that provides facilities management services on a fee-for-service basis as its core business is referred to as a *facilities management contractor* to distinguish it from an in-house facilities management department. *Facilities management function* is also used to refer in a generic, collective sense, to the operational group of facilities management resources delivering facilities management services to the organization.

The term *strategy* also appears numerous times throughout the book. 'An organization's strategy describes how it intends to create value for its shareholders, customers and citizens' (Kaplan and Norton 2004, p. 4). *Business strategy* can be defined as 'a specific description of how a business unit is to compete in its markets' (Hamermesh 1986, p. 34). In the context of this book, *business strategy* refers to the method that a business or organization adopts as its means of delivering its planned outputs and outcomes in order to achieve its goals. It includes how all the relevant processes, policies, resources and organization structures will be deployed to achieve the stated goals. *Strategy*, as it is used in a facilities management context, such as *maintenance strategy*, refers to the specific technical and management approach (including all the relevant processes, policies, resources and organization structures) adopted to achieve specific objectives.

Acquisition is generally used to refer to the transfer of ownership or control of a particular facility or asset to the organization. *Procurement* is used to refer to the transactional processes implemented to effect the acquisition. The term *system* has been used in basically two contextual aspects. In the context of this book the term *system* can refer to a framework of policies, processes and procedures designed to achieve a specific business objective or set of objectives. It can also refer to the integrated configuration of hardware, software, procedures, inputs and outputs that make up a management support tool (e.g. computerized maintenance management system) or building services control and management tools (e.g. lift control system) or a facilities management system.

1.7 Book structure

We firmly believe that current practices in facilities asset management in many organizations are seen as an adjunct to business management, leading to the facilities functions being seen as essentially reactive responses to the impact of business dynamics. Our contention is that facilities asset management must be business-driven as an essential part of prudent management of business resources. The justification for this contention forms the basis of the structure of this book.

This book will take the reader through the principal components of facilities asset management which reflect the life-cycle stages of a facility asset. Each of these principal components has key process elements. A process element of facilities asset management is defined in the context of this book as a management or technical process undertaken for a specific facilities asset management objective. The outcomes of such processes have implications for business in terms of capital investment, space capacity, operational reliability, health and safety, operating and maintenance costs and other issues such as environmental sustainability.

Key process elements of each of the principal components of facilities asset management are then discussed. Finally, the book will discuss the organizational implications of facilities asset management.

References

Alexander, K. Ed. *Facilities Management – Theory and Practice.* E & FN Spon. 1998.

Alexander, K., Atkin, B., Brochner, J. and Haugen, T. Eds. *Facilities Management Innovation and Performance.* Spon Press. 2004.

Apgar, M. 'Managing Real Estate to Build Value'. *Harvard Business Review.* Nov–Dec 1995, pp. 162–179.

Apgar, M., IV and Bell, M. 'Managing Real Estate to Leverage Change: The Dun & Bradstreet Case'. *Site Selection.* Dec 1995, pp. 935–38.

Arthur Anderson *Real Estate in the Corporation: The Bottom Line from Senior Management.* 1993.

Atkin, B. and Brooks, A. *Total Facilities Management.* 3rd Edition. Blackwell Publishing. 2009. (ISBN: 978-1-4051-8659-9).

Avis, M., Gibson, V. and Watts, J. *Managing Operational Property Assets*. GTI. 1989.

Barrett, P. and Baldry, D. *Facilities Management: Towards Best Practice.* Blackwell Science. 2003.

Becker, F. *The Total Workplace.* Van Nostrand Reinhold. 1990.

Becker, F. and Steele, F. *Workplace by Design: Mapping the High-Performance Workscape.* Jossey-Bass. 1994.

——*Offices at Work: Uncommon Strategies that Add Value and Improve Performance.* Jossey-Bass. 2004.

Best, N., de Valence, G. and Langston, C. Eds. *Workplace Strategies and Facilities Management: Building in Value.* Butterworth-Heinemann. 2003.

Booty, F. *Facilities Management Handbook.* 4th Edition. Butterworth-Heinemann. 2009.

Boussabaine, A. and Kirkham, R. Whole *Life-cycle Costing: Risk and Risk Responses.* Blackwell Publishing. 2004 (ISBN 1-4051-0786-3).

Brand, S. *How Buildings Learn – What Happens after They're Built?* Viking Press. 1994. (ISBN 0-670-83515-3).

Cooperative Research Centre for Construction Innovation (CRC). *FM as a Business Enabler.* 2007.

CoreNet Global. *Corporate Real Estate 2010 – Integrated Resource and Infrastructure Solutions.* 2004.

Cotts, D. and Rondeau, E. *The Facility Manager's Guide to Finance and Budgeting.* AMACOM. 2003. (ISBN: 0-8144-0562-2).

Cotts, D., Roper, K. and Payant, R. P. *The Facility Handbook.* 3rd Edition. AMACOM. 2010. (ISBN: 987-0-8144-1380-7).

Federal Highway Administration (FHWA). *Asset Management Primer.* Washington, D.C. Department of Transportation. 1999.

Finch, E. *Facilities Change Management.* Wiley-Blackwell. 2012.

Gallup. *Shaping the Workplace for Profit.* Commissioned by Workplace Management. 1996.

Hamermesh, R. *How Senior Managers Produce Results*, John Wiley and Sons Inc., Toronto. 1986.

Hastings, N. A. J. *Physical Asset Management.* Springer. 2010. (ISBN: 978-1-84882-70-9).

Haynes, B. and Nunnington, N. *Corporate Real Estate Asset Management: Strategy and Implementation.* EG Books. 2010. (ISBN: 978-0-7282-0573-4).

Hinks, J. and Reuvid, J. *Managing Business Support Services.* 2nd Edition. Kogan Page. 2002. (IBN:0-7479-3900-9).

i-FM.net. *Facilities Management: Realizing the Value of the Investment.* 2011.

Jorroff, M., Louargand, M., Lambert, S. and Becker, F. *Strategic Management of the Fifth Resource: Corporate Real Estate.* The Industrial Development Research Council, U.S.A. 1993.

Kaplan, R. and Norton, D. *Strategy Maps – Converting Intangible Assets into Tangible Outcomes.* Harvard Business School Publishing Corporation, Massachusetts. 2004.

Langston, C. *Life-cost Approach to Building Evaluation.* Elsevier. 2005 (ISBN: 0-75066-630-7).

Langston, C. and Lauge-Kristensen, R. *Strategic Management of Built Facilities.* Butterworth-Heinemann. 2002.

Lindholm, A., Gibler, K.M. and Leväinen, K.L. 'Modeling the Value-adding Attributes of Real Estate to the Wealth Maximization of the Firm'. *The Journal of Real Estate Research.* 2006. Vol. 28, No. 4, pp. 445–75.

Lloyd, B. 'An Effective Property Strategy – The Key Challenge for the 1990s'. *Facilities.* 1992. Vol. 10, No. 12, pp. 9–12.

Lloyd, C. *Asset Management – Whole-life Management of Physical Assets.* Thomas Telford. 2010.

McGregor, W. and Then, D. S.-S. *Facilities Management and the Business of Space.* Arnold. 1999; Butterworth-Heinemann. 2001.

National Research Council (NRC). *Stewardship of Federal Facilities: A Proactive Strategy for Managing the Nation's Public Assets.* Committee to Assess Techniques for Developing Maintenance and Repair Budgets for Federal Facilities. The National Academies Press. 1998.

——*Investments in Federal Facilities: Asset Management Strategies for the 21st Century.* The National Academies Press. 2004.

Nutt, B. 'Infrastructure Resources: Forging Alignments between Supply and Demand'. *Facilities.* 2004. Vol. 22, Nos. 13/14, pp. 335–43.

Oschrin, J. 'The Life-Cycle Audit: A Way to Understand and Communicate the Need for Facility Reinvestment'. *Journal of Facilities Management.* 2004. Vol. 3, No. 3, pp. 284–92.

Sracel, H. 'Aligning Corporate Real Estate with Company Goals'. *Facilities Design & Management.* Nov 1994, pp. 48–53.

Then, D. S.-S. 'Facilities Management – The Relationship between Business and Property'. *Proceedings of EURO FM/FMA Conference on Facility Management European Opportunities*, Brussels. May 1994, pp. 253–62.

——'Integrated Resources Management Structure for Facilities Provision and Management'. *Journal of Performance of Constructed Facilities.* Feb 2003. Vol. 17, No. 3, pp. 34–42.

Then, D. S.-S. and Tan, T. H. 'Real Estate and Facilities Management Alignment to Business Needs – A Conceptual Model'. *Proceedings of 2010 International Conference on Construction and Real Estate Management.* Dec 1–3, 2010. Brisbane, Australia. Vol. 1, pp. 33–37.

Theriault, M. *Managing Facilities and Real Estate.* Woodstone Press. 2010.

Varcoe, B. J. 'Premises of Value'. *Facilities.* 1992. Vol. 10, No. 3, pp. 14–20.

Veale, P.R. 'Managing Corporate Real Estate Assets: Current Executive Attitudes and Prospects for an Emergent Management Discipline'. *The Journal of Real Estate Research.* Fall 1989. Vol. 4, No. 3, pp. 1–22.

Zeckhauser, S. and Silverman, R. 'Rediscover Your Company's Real Estate'. Harvard Business Review. Jan–Feb 1983, pp. 111–17.

Appendix 1: Definitions of facility/facilities management by professional institutions

International Facility Management Association (IFMA)

> Facility management is a profession that encompasses multiple disciplines to ensure functionality of the built environment by integrating people, process, place and technology.
>
> www.ifma.org/resources/what-is-fm/default.htm

British Institute of Facilities Management (BIFM)

> Facilities management is the integration of processes within an organisation to maintain and develop the agreed services which support and improve the effectiveness of its primary activities.
>
> BIFM has formally adopted the definition of facilities management provided by CEN, the European Committee for Standardisation and ratified by BSI British Standards.
>
> www.bifm.org.uk/bifm/about/facilities

Note: Before the adoption of the CEN definition, BIFM used the following definition:

> Facilities Management is the integration of multi-disciplinary activities within the built environment and the management of their impact upon people and the workplace.

Facility Management Association of Australia (FMA)

> A business practice that optimises people, process, assets and the work environment to support the delivery of an organisation's business objectives.

Note: The above definition has been withdrawn from the FMA website (www.fma.com.au) with the following explanation:

> To alleviate the issues caused by numerous different local and international definitions of facilities management, FMA Australia is currently working with Standards Australia, the International Standards Organisation, along with our members and international partner organisations to create a single definition (or consistent series thereof) which will effectively communicate the profession and industry to informed and uninformed stakeholders alike.
>
> www.fma.com.au/cms/index.php?option=content&task=view&id=45 (accessed 25 May 2012)

Hong Kong Institute of Facility Management (HKIFM)

> Facilities Management is the integration of multi-disciplinary activities within the built environment and the management of their impact upon people and the workplace.
>
> www.HKIFM.org.hk

Japan Facility Management Association (JFMA)

> It is a comprehensive management approach for the optimization of the ownership, utilization, operation, and maintenance of the business real properties (land, buildings, structures, equipment, etc.) … so that they could contribute to the overall management of the business.
>
> www.jfma.or.jp/en/whatsFM/index.html

2 Facilities and business

There is a clear trend, in recent years, that supports the proposition that the push by businesses to demonstrate value for money necessitates a management approach that embraces all of a business's resources and emphasizes effectiveness, on top of general economic efficiency (Jensen 2011, Jensen *et al.* 2012, Krumm and de Vries 2003, Then 1997). The economic imperatives imposed by global competition and the continuous drive for improved performance have promoted a need to consider the implications of strategic business decisions on the corporate facilities' asset base. Such implications include the following:

- Businesses increasingly demand flexibility in response to competition in order to survive. The supporting physical infrastructure (the facilities portfolio) must be able to respond accordingly, necessitating a proactive management approach based on full knowledge of the portfolio (that is, opportunities, as well as constraints).
- Increasingly, the arena for conducting businesses no longer respects national boundaries. The economics of international comparative advantage dictate site selection, with major implications for facilities provision beyond just production capacity, including associated social infrastructure to house the local labour.
- The rapid and widespread emergence of e-commerce as an effective way to conduct business.
- Buildings, as operational assets, and the space that business units occupy are a legitimate expense and no longer regarded as 'free' goods hidden within the corporate overheads.
- Buildings, as physical assets, are durable assets demanding a life-cycle approach in their management over time.
- The service element of the delivery of facilities-related services (customer interface) is seen as a critical dimension of the facilities-provision and facilities-support services delivery package.
- The adoption of environmentally responsible business policies is having a flow-on effect on the environmental performance of existing and new facilities.

The above dynamics are impacting upon the way in which businesses need their people to work in terms of location, processes and tools. Yet, for many organizations, they are ponderously slow to respond to such dynamics in terms of the work environment of their employees, creating situations where their facility asset base is inhibiting corporate development (McGregor 2000).

2.1 Business context of facilities

Buildings as a class of durable facility assets take on many shapes and forms, but are fundamentally real estate assets which appear on an organization's financial statements. Therefore the first implication of facility assets for any organization owning or leasing them is the financial costs of ownership and occupancy. These include leasing costs, depreciation, capital investment and operating costs. They are, arguably, the second highest business cost after salaries (Varcoe 1993, Steiner 2005). The second implication is the impact of facilities on the business of the organization in terms of providing the working environment in which business is conducted and where production processes occur (Then 2004, Varcoe 1993).

Increasingly, organizations are recognizing the management of facility assets (real estate) as a business function (e.g. Apgar 1995; Avis, Gibson and Watts 1989, Then 1997, 1998; Varcoe 1993). This is driven by the realization that the impact of facilities is not only on costs but also on staff and process productivity, recruitment and retention, and competitive advantage (e.g. customer service, location and perceptions of quality service). By focusing on the value contribution of facilities, organizations can better align these resources with business strategies and objectives (Jensen 2010; Jensen, *et al.* 2012).

Consideration of the value contribution of facilities to business must include an understanding of the nature of the business and the degree to which the facilities are designed, constructed and serviced to meet the specific requirements of the business and its processes. This enables the management effort and emphasis on the management of facilities to be put in the proper perspective. Facilities may range from the very highly specialized, such as high technology manufacturing facilities, research laboratories and tertiary hospitals, to the different grades of general office accommodation and ancillary facilities for storage and other support purposes. Facilities can also be held as property investments to provide a return on investment as a primary business objective of ownership (e.g. a real estate investment trust). This latter class of facilities has a different business and operational relationship with the owner compared with the occupiers who lease the facilities. The management perspective of the owners is primarily a return on investment, while the end users' perspective is that of a customized workspace and service delivery environment. In both cases, however, the facilities represent a business resource with revenue and cost implications.

The business purpose and complexity of facilities have a significant impact on the facility solutions available to the organization and the demand for attention from senior management. The development of facility solutions (e.g. location, design, construction, acquisition, operation, maintenance, management and disposal) becomes more critical and demanding as the business processes undertaken within a facility increase in technical complexity, specialization and criticality to the business objective (see Figure 2.1).

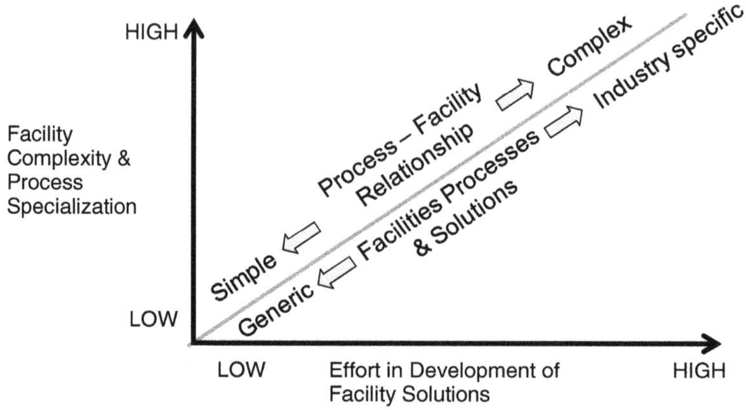

Figure 2.1 Facilities complexity

Facility decisions at the more complex end of the spectrum of facility specialization are more likely to have greater technical, financial and business implications and hence require more robust planning and decision-making processes supported by sound evaluations of the design options and subsequent in-use phase of the facility assets. For example, the design and construction of hospitals in Australia are guided (amongst other things) by a comprehensive set of national guidelines known as the Australasian Health Facility Guidelines (http://healthfacilityguidelines.com.au). Evidence-based design is also used in the health care sector to increase the likelihood that the significant financial investments in new health care facilities deliver the desired outcomes and reduce the risks of a mismatch between functional requirements and technical design (Becker and Parsons 2007). In contrast, the same degree of comprehensiveness does not exist for general office space used by the public sector in Australia. More recently, design considerations of commercial offices have taken on a new emphasis as a result of the inevitable impact of rapid technological developments in computing, communication and social networking. Features of modern offices include facilities that support mobility and flexibility for users who choose where and when they wish to work (Meel, Martens and Ree 2010, Then 2010).

The above highlights the importance of business managers developing competent qualitative and quantitative knowledge of the relationship between their business processes and the physical environment in which they are performed. This knowledge underpins the development of functional requirements and performance standards which are fundamental to the design, construction and configuration of facilities that meet business requirements and are sustainable in terms of operating and maintenance costs.

Whilst the above seems obvious, research has revealed that the crucial link between business and facilities is often not fully understood or appreciated (e.g. RICS 2002, Then 1997, Bannock & Partners 1994, Morris 1993, Veale 1989; Zeckhauser and Silverman 1981). However, in recent years there are signs of growing awareness of the vital role of facilities in supporting the achievement of corporate objectives (e.g. Apgar 2009, Varcoe 2000, Then 2000).

In the main, the current practice of facilities management remains entrenched in technical and transaction management. Its focus is on the operational management of maintenance, leasing, engineering services, design, fit-out and refurbishment, and construction – to name a few facilities support services. Most facilities management models are still internalized operational models based essentially on supply management, whereas demand management is the motivator of organizational dynamics. In their effort to address the demand–supply imbalance perception of facilities management practice, Then and McEwan (2007) provided an economic model that illustrates the fundamental relationships between business need for facilities (i.e. business–facility interface) and the relationship between facilities management goods and services resulting from the utilization of a facility (i.e. facility–FM goods and services interface). Figure 2.2 illustrates the economic fundamentals of the need to balance the cost of facilities provision with the benefits derived, as an FM resource consumption model (Then and McEwan 2007).

It is the management dimension of facilities management that remains, arguably, undeveloped. In the main, the facility and facility services relationship is well understood by facilities managers. However it is the business-facility interface or relationship which requires management knowledge and skills that are currently poorly developed.

The proposed FM resource consumption model places the business context of Supply and Demand Management at the heart of understanding the economics of facilities management. The model illustrates the context and relationship between business and its facility needs, and between a functional facility and its required services. It is important to stress the demand–supply cycle between each of three components (i.e. Business – Facility – FM Goods and Services). It is the demand

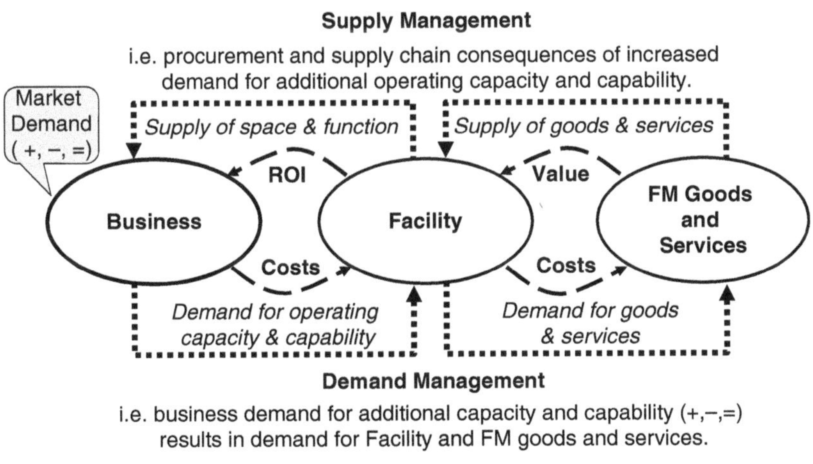

Figure 2.2 A facilities management resource consumption model

for operating capacity and capability that leads to the supply of space and function. Correspondingly, it is the utilization of the facility by the business that generates the demand for goods and services within the functional facility that leads to their supply. In terms of business resource management, the focus of Demand Management is to minimize operating and occupancy costs, while the focus of Supply Management is effective procurement and supply chain optimization.

In summary, the FM resource consumption model takes into account how a facility can contribute to maximizing the wealth and value of a business through the tangible and intangible benefits that it can derive through prudent facilities management.

2.2 Facilities–business relationships

As noted above, business processes drive the realization of facilities solutions, employing facilities management processes with varying degrees of sophistication, depending on the nature and complexity of business processes. However, the relationship of an organization with its facilities is complex and multidimensional and extends over the life of the facilities or at least the duration of ownership and use. A clearer understanding of this dynamic relationship is needed as a first step along the journey towards effective management of facilities as a business resource and their alignment with business requirements (Hinks 2000, Osgood 2003).

Table 2.1 provides a very brief overview of some of the interrelationships between the facilities dimensions in terms of the life-cycle stages of a facility asset and six key aspects of the business dimension common to most business organizations. The six aspects are by no means comprehensive, but have been chosen for illustrative purposes only. The interrelationships have been presented for each stage of a facility asset life cycle. The six business aspects of an organization are:

- the business organization operating in the facilities in terms of business strategy, processes, policies, outputs, customers;
- the people working in the space within the facility such as staff, visitors, service providers, community;
- the technology supporting the organization, including building engineering systems, business information technology systems, facility management systems, process technology;
- the financial factors such as capital investment, operating costs (business and facilities) and asset depreciation;
- the business partners, including suppliers, contractors, collaborative partners; and
- the risks to business in terms of legal compliance, finance, safety, business and environmental risks.

The business–facilities interface is multidimensional and more complex than is possible to be portrayed by a simple matrix. For example, facilities technology affects the occupants (people), and facilities risks are also related to maintenance costs. However, the matrix (Table 2.1) is meant only to demonstrate some of the

Table 2.1 Facilities assets and business interrelationships

Business	*Facilities life-cycle stages*			
	Planning	*Acquisition*	*Utilization and Maintenance*	*Renewal/Disposal*
Organization (business strategy, processes, policies, outputs, customers)	Strategic directions, business trends and requirements drive facilities demand and performance standards.	Policies on design, procurement, leasing and construction project management influence facilities acquisition processes and outcomes.	Space utilization, condition of facilities and operational reliability affect business processes and customer perceptions.	Renewal and disposal of facilities provide opportunities for improving alignment of facilities with business strategy and processes.
People (staff, visitors, service providers, community)	Expectations of staff, visitors, service providers and community influence planning considerations and performance standards.	Staff, visitors, service providers and community can provide valuable input into design and construction of facilities.	Facilities users affect wear and tear of facilities. Condition of facilities and operational reliability can affect productivity and pose risks to users.	Disruptions due to renewal and disposal affect facilities users and need to be managed sensitively and effectively.
Technology (building, business process, facilities and business management)	Strategic directions, business trends and requirements drive the technology requirements to be considered in planning.	Design and construction technologies enhance efficiency of facilities acquisition and future management.	Building technologies influence workspace usability. FM systems enable and enhance facilities maintenance and workspace management.	Changing technology may drive renewal and disposal due to obsolescence and mismatch with business requirements.
Financial (capital, expense, depreciation)	Budgets influence planning outcomes as well as future operating costs and performance of facilities.	Budgets influence acquisition outcomes. Acquisition processes affect budgets (initial capital and future operating or lease expenses).	Operating and maintenance costs impact on budgets for recurrent expenditure. Underfunding of maintenance has long-term consequences.	Capital budgets for renewal need to be planned for. Disposal of facilities unlock capital and could produce 'windfall' revenues.

Table 2.1 (continued)

Business	*Facilities life-cycle stages*			
	Planning	*Acquisition*	*Utilization and Maintenance*	*Renewal/Disposal*
Partners (suppliers, contractors, collaborative partners)	Facility location, logistics and communications are key planning considerations. Opportunities to optimize geographical factors are greatest at planning stage.	Opportunities to optimize geographical factors diminish at acquisition stage. However, globalization and technology have introduced new dimensions in business operations.	Costs of doing business with suppliers, contractors and collaborative partners will manifest at this stage and affect the bottom line.	Opportunities to review costs of doing business with suppliers, contractors and collaborative partners and implications of facilities location, technology and strategy.
Risk (legal compliance, financial, safety, business, environmental)	Opportunities to address various types of risks. Management of planning process risks and identifying risks associated with planning outcomes.	Implementation of risk management features through design and construction. Management of acquisition process risks (governance, financial, contractual).	Risks of occupancy and use throughout the life of the facilities. Managing risks to business continuity and other long-term liabilities (e.g. deferred maintenance).	Renewal and disposal process risks (governance, financial, contractual). Environmental impact of demolition and hazardous substances (e.g. asbestos).

key implications across the facilities life-cycle stages on selected business aspects and to emphasize the point that a more comprehensive understanding of the business–facilities relationship will enhance management, planning, design, construction, maintenance and operation of facilities resources.

One of the most important relationships between facilities and business is their impact on the balance sheet in terms of financial value and as an expense item in statements of financial performance. Facilities assets are capital-intensive and appear on the balance sheet as 'capital assets' (that is, with a life of more than a year and not bought and sold in the course of normal business). Their values are depreciated as recognition of the cost of assets in delivering business outputs, and the depreciation amounts are listed as a non-cash expense in statements of financial performance. Other expenses like operating, maintenance and utilities costs appear in statements of financial performance. The financial recognition of facilities as assets on the balance sheet forms part of the total value of the business to the shareholder and potential investor in the case of private sector organizations (or tax-payer in terms of public sector organizations). The operating and maintenance cost of facilities as an expense is an important consideration for both private and public sector organizations dealing with issues of market competitiveness and funding constraints, respectively.

From an asset and financial management perspective, three key issues need to be considered:

- the provision for replacement of the asset components as they become due (they are subject to wear and tear, obsolescence and changes in functional requirements);
- the potential for better use of the capital invested in the assets should they be sold and alternative facility solutions be adopted (e.g. leased facilities) and the implications thereof on business operations;
- capital budgeting for new facilities to meet forecast future requirements so that these can be funded and delivered at the time required.

As noted in Chapter 1, facilities assets have components of varying longevities, and their replacement becomes problematic if the necessary financial capability does not exist or is inadequate when they (or their major components such as an air conditioning plant) become due for replacement. Depreciation is a non-cash expense and is not necessarily dedicated to the future replacement costs of assets consumed in the course of business. Therefore the importance of life-cycle replacement planning as part of the budgetary process should not be understated and in fact should be a vital part of strategic facilities planning and maintenance. A planned capital replacement programme maintains the value of the facilities portfolio and its capability to support business functions effectively over its useful life.

A 'sale and lease-back' facilities solution could be a viable proposition for some organizations in order to release the capital invested in facilities assets for other purposes. The assets no longer appear on the balance sheet, but their full costs are now reflected as a predictable lease expense on statements of financial performance. The facilities could be owned by a real estate investment trust or other entity as an income-earning asset. While this is one possible scenario for asset ownership, the important point highlighted is the potential of other asset ownership models as a means of releasing capital tied down in facilities assets.

Facilities take time to plan and acquire or build. They require significant capital funds to be planned for and set aside in order to be ready in time for the delivery of increased capacity and services. The timing of facilities being completed or becoming available on time to meet the demand for space is crucial for both the private (e.g. to avoid loss of business opportunities) and public sectors (e.g. to avoid community criticism). Capital budgeting for new facilities assets based on sound strategic facilities planning enhances the ability to meet new business commitments in a timely and economical manner.

2.3 Facilities as an undermanaged resource

As a general industry trend the above ideal is not always reflected in practice, and facilities remain an undermanaged resource. Facilities management is not yet fully integrated as a business function and continues to be managed as an administrative and technical function (Apgar 1993, 2009; Then 1998, 2000; Bannock & Partners 1994, Debenham Tewson Research 1992, Avis, Gibson and Watts 1989). Given the increasing corporate acknowledgement of the role of facilities in recruitment, retention and productivity of staff, the focus on costs of facilities must now be balanced by a more holistic evaluation approach that justifies not only costs incurred but also other facilities management initiatives that add value to the core business in terms of customer and shareholder or stakeholder value (Jensen 2010, Gallup 1996, Henley Centre 1996, Apgar 1995). Input measures such as cost per unit space would then be complemented by output measures such as revenue per unit space or service outputs per unit space in the case of public sector organizations (Steiner 2005, Then and Tan 2004, Varcoe 2002). This would provide the catalyst for better integration of facilities management with other aspects of business management. Given the interrelationships with other parts and systems of the organization, enhancements of the various relationship interfaces would initiate and encourage greater integration of facilities management with the rest of the organization and bring greater recognition of facilities management as an essential business function (see Figure 2.3).

The benefits of greater integration of facilities dimensions with business planning include:

- better facilities solutions as a result of greater organizational awareness of facilities as a business resource and their value contribution;
- improved potential for aligning systems, resources and facilities with business needs through active involvement of all stakeholders in facilities management;
- enhancing the professionalism and service focus of facilities management through better integration with business strategies and goals.

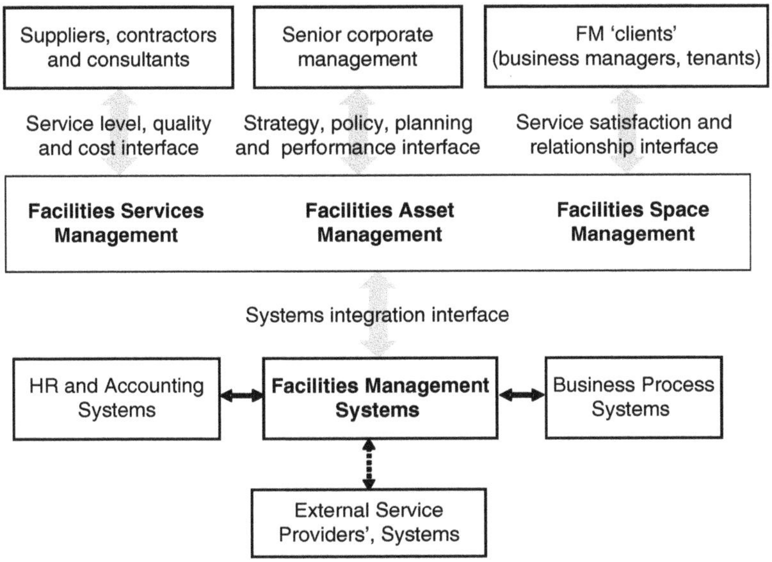

Figure 2.3 Facilities management – organizational interfaces

References

Anderson, A. 'Wasted Assets? A Survey of Corporate Real Estate in Europe'. 1995. Arthur Anderson.

Apgar, M. 'Managing Real Estate to Build Value'. *Harvard Business Review*. Nov–Dec 1995, pp. 162–79.

——'Uncovering Your Hidden Occupancy Costs'. *Harvard Business Review*. Nov 2003, pp. 124–36.

——'What Every Leader Should Know about Real Estate'. *Harvard Business Review*, Nov 2009, pp. 100–07.

Avis, M., Gibson, V. and Watts, J. *Managing Operational Property Assets*. GTI. 1989.

Bannock & Partners Ltd. *Property in the Board Room – A New Perspective*. Commissioned by Hillier Parker. 1994.

Becker, F. and Parsons, K. 'Hospital Facilities and the Role of Evidence-based Design'. *Journal of Facilities Management*. 2007. Vol. 5, No. 4, pp. 263–74.

Debenham Tewson Research. *The Role of Property – Managing Cost and Releasing Value*. 1992.

Gallup. *Shaping the Workplace for Profit*. 1996.

Henley Centre. *The Miliken Report: Space Futures*. 1996.

Hinks, J. 'Business Strategy – Is FM a Core Concern?' *Facilities*. Guest editorial. 2000. Vol. 18, No. 10/11/12.

Jensen, P. A. 'The Facilities Management Value Map: A Conceptual Framework'. *Facilities*. 2010. Vol. 28, Nos. 3/4, pp. 175–88.

——'Organisation of Facilities Management in Relation to Core Business'. *Journal of Facilities Management*. 2011. Vol. 9, No. 2, pp. 78–95.

Jensen, P. A., Voordt, T., Coenen, C., Fellen, D. and Lindholm, A. 'In Search for the Added Value of FM: What We Know and What We Need to Learn'. *Facilities.* 2012. Vol. 30, Nos. 5/6, pp. 99–217.

Krumm, P. J. M. M. and de Vries, J. 'Value Creation through the Management of Corporate Real Estate'. *Journal of Property Investment & Finance.* 2003. Vol. 21, No. 1, pp. 61–72.

McGregor, W. 'Preparing for an Uncertain Future'. *Facilities.* 2000. Vol. 18, Nos. 10/11/12, pp. 402–10.

Meel, J., Martens, Y. and Ree, H. J. *Planning Office Spaces – A Practical Guide for Managers and Designers.* Centre for Facilities Management – Realdania Research, Techical University of Denmark. 2010.

Morris, J. 'The Gentle Art of Reframing: Making an Impact on the Senior Management Agenda'. *Facilities.* 1993. Vol. 11, No. 12, pp. 16–18.

Osgood, R. T. 'Strategic Alignment: Measuring Real Estate in the Context of Organizational Outcomes'. *Leader.* May 2003, pp. 16–18.

RICS. *Property in Business – A Waste of Space?* 2002.

Steiner, J. 'The Art of Space Management'. *Journal of Facilities Management.* 2005. Vol. 4, No. 1, pp. 6–22.

Then, D. S.-S. 'A Model for Considering Operational Property as an Enabling Resource to Business'. *Proceedings of RICS Cutting Edge Property Research Conference*, Dublin. 1997, pp. 600–11.

——'Corporate Leadership in Real Estate and Facilities Management'. Facilities Management and Maintenance: the way ahead into the Millenium. Quah, L. K. (Ed.) McGraw-Hill. 1998, pp. A9–A16.

——'The Role of Real Estate Assets in Supporting the Fulfillment of Corporate Business Plans – Key Organizational Variables for an Integrated Resource Management Framework'. *Facilities, 2000.* Vol. 18, Nos. 7/8, pp. 273–80.

——'Concepts in Facilities Management'. *Architect's Handbook of Construction Project Management.* Murray, M. and Langford, D. (Eds) RIBA Enterprises. 2004, pp. 297–334.

——'The Business of Space'. *Facilities Change Management.* Finch, E. Ed. Blackwell Publishing. 2010, pp. 57–75.

Then, D. S.-S. and McEwan, A. 'A Facilities Management Resource Consumption Model'. *Proceedings of CIB World Building Congress 2007 – Construction for Development.* Cape Town, South Africa. 21–25 May 2007, pp. 600–11.

Then, D. S.-S, and Tan, T. H., 'Assessing Building Performance – An Integrated Model'. CD *Proceedings of CIB World Building Congress 2004 – Building for the Future*, Session CIB T5S1. Toronto, Canada. May 2004, pp. 1–9.

Varcoe, B. J. 'Facilities: The Business Dimension'. *Facilities Management.* Dec 1993, pp. 5–7.

——'Implications for Facility Management of the Changing Business'. *Facilities.* 2000. Vol. 18, Nos. 10/11/12, pp. 383–91.

——'The Performance Measurement of Corporate Real Estate Portfolio Management'. *Journal of Facilities Management.* Aug 2002. Vol. 1, No. 2, pp. 117–30.

Veale, P. R., 'Managing Corporate Real Estate Assets: Current Executive Attitudes and Prospect for an Emergent Management Discipline'. *Journal of Real Estate Research.* 1989. Vol. 4, No. 3. pp. 1–22.

Zeckhauser, S. and Silverman, R. 'Rediscover Your Company's Real Estate'. *Harvard Business Review.* 1981, Jan–Feb. Vol. 61, Issue 1, pp. 111–17.

3 Facilities asset management

The pervasive impact of technology, the rise of the service culture and the search for greater efficiency in the use of all resources have challenged the real estate and facilities industry to deliver new and more responsive facilities solutions to meet the needs of the occupiers, customers and a wide range of other stakeholders. Land and buildings are probably the slowest of all the strategic resources (finance, people, technology, and land and buildings) to respond to change. This is partly attributable to the nature of the product, which is bound by legal property rights, regulatory controls of planning, construction and occupation involving a large financial outlay and the operations of the real estate (property) market. As a consequence of the long lead-in times for asset creation and asset transactions (i.e. non-liquid assets) there is a need for organizations owning or leasing facilities assets to plan changes in a very systematic way. Thus, there is a need for a structured approach to the management of ongoing operational aspects and the long-term change in facilities assets in order to ensure their continuous fitness for purpose to the business.

3.1 Facilities as a business resource

Facilities provide the business environment for staff, customers, products, services, processes, production equipment and supplies. Globally, there is growing awareness of the potential for facilities to enhance competitive advantage through their image, location, functionality, technology, amenities and attractiveness to staff as workplaces of preference (Leaman 1991a, 1991b, Hinks 2000, Krumm 2003, Price *et al.* 2003, Tay 2006, Jensen 2011). Changing the paradigm of the perception of facilities as cost centres to strategic resources capable of enhancing business outcomes has the potential to significantly benefit organizations in the private and public sectors. Fortunately a more enlightened view is emerging across industry generally that sees facilities as a valuable resource providing work and business environments that very much impact on the bottom line and future long-term viability and success. This enlightened view also recognizes the far wider role of facilities management in three key aspects – workspace management, asset management and facilities services management.

Including facilities as a business resource with the three other key business resources (people, finance and technology) enhances the business platform for

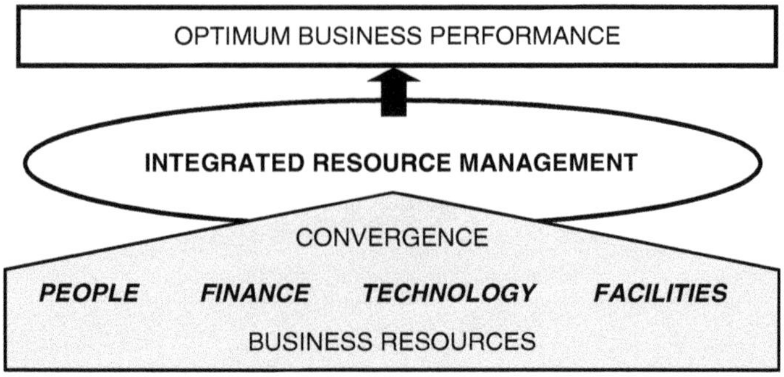

Figure 3.1 Strategic business resources

high performance and achievement of strategic direction (Varcoe 1993, Robertson 2000). The platform is balanced (four 'legs' instead of three) and facilitates integrated management of these four business resources to achieve optimum business performance (Figure 3.1). Such an integrated resource management view has been gaining momentum in recent years in real estate and facilities management literature (Kincaid 1994, Then 2000, Nutt 2004).

3.2 Management challenges

Managing facilities as a business resource presents facilities and business managers with a range of challenges. These include issues such as their role in supporting the business, the physical and financial implications of ownership and use, legal compliance, business continuity and risk. These challenges drive the need for not only a robust and professional facilities management capability but also a keen business focus on facilities as a resource and how they can best be leveraged towards achieving business objectives (Jones 1995, Loch 2000, Roberts 2001, Kaya *et al.* 2004). The challenges outlined below are by no means exhaustive, but are intended to be illustrative only for the purpose of highlighting the need for effective facilities asset management.

3.2.1 Business demand for space

Central to the idea of facilities as a business platform is the extent to which the facilities meet the needs of business in terms of factors such as location, space, functionality, technology, appearance, condition, security, supporting infrastructure and access, cost and, increasingly, environmental sustainability features (McGregor and Then 2001). The challenges to business include the following:

- the ability to assess the performance of facilities in meeting current and future business requirements;
- dealing with issues of under- and over-capacity and alignment with business needs;

- maintaining strategic spare capacity in anticipation of future needs;
- adjustments to facilities in response to changes in business directions and demand for space;
- dealing with options in facilities procurement models (e.g. direct ownership, lease, sale and lease-back, public-private partnerships, etc.);
- facilities as a cost to business (financial performance) and as long-term physical assets (balance sheet).

3.2.2 Managing the assets

Being real estate assets, building facilities are long-life assets representing in the majority of cases a long-term investment, with a long trail of ongoing maintenance and operating costs over their entire life cycle. Renewal, operating and maintenance costs over the life of the facilities have been estimated to be more than five times their initial capital value. These costs impact on business operating costs, capital requirements and the cost of borrowings, all of which are reflected in the balance sheet and statements of financial performance. More importantly, they impact on the cost of business processes.

Over time, as facilities decline in condition through usage their designed capacity and functionality may also diverge from operational requirements as a result of changes in occupancy, processes and technology; maintenance costs increase and capital costs of renewal loom as significant factors in financial planning. These renewal cycles are also likely to become shorter due to rapid changes in technology, dynamic global economic conditions and the underlying need for business to remain competitive and respond quickly to new opportunities and threats. Lack of planning and an appropriate response to the decline in physical condition and functionality could lead to these assets becoming liabilities, expose the business to risks and adversely affect the efficiency and effectiveness of business processes (Lynch 2001a, 2001b). The challenges to business include the following:

- the capacity to manage and maintain the facilities from planning through procurement, utilization, renewal and eventual disposal;
- the allocation of appropriate levels of capital and operational funding to sustain the facilities to meet business requirements and expectations;
- the ability to balance the initial costs of facilities with the long-term operating and maintenance costs to achieve the most economical life-cycle cost of facilities;
- the choice of the most appropriate model for facilities management and the delivery of facilities support services;
- managing within the parameters and dynamics of the facilities management and construction industries' business environments to achieve the best outcomes under varying economic scenarios.

3.2.3 Compliance with statutory requirements

In many countries building regulations are continually being updated, improved and more rigorously enforced as a result of concerns about health and safety. They add significant complexity to the tasks of owning, operating and managing building facilities. Building failures, devastating fires and building- and construction-related accidents as well as contamination issues are driving a hardening of community attitudes towards the need for more tightly regulated facilities. The implications for building facility owners are greater exposure to legal action for non-compliance, costs of alterations and upgrading to meet new regulations and higher maintenance and operating costs, as well as the costs for training, safety education and safety equipment. As a consequence there is a greater need for more professional facilities management and greater pressure on business managers to focus on the safety and regulatory compliance of their facilities in order to avoid incidents and their consequences. The challenges for facilities and business managers include the following:

- maintaining up-to-date knowledge of the laws and regulations pertaining to their facilities and particular industry;
- having the capacity to monitor and implement statutory and safety requirements as well as an effective strategy and process for responding in a timely fashion to any breaches or potential breaches;
- maintaining a balance between risks, costs and benefits in implementation of measures beyond the minimum compliance standards;
- establishing policies and standards in relation to statutory compliance issues and a framework for effective implementation.

3.2.4 Business continuity

Climate change, natural disasters, mass demonstrations and the evolving threats of religious and politically motivated violence against people and facilities have focused attention on facilities as strategic assets and the impact on business continuity should these facilities be disabled or isolated. Large-scale and devastating events around the world have tragically shown that the paradigm of traditional emergency preparedness measures has been irrevocably changed. Facilities and business managers need to develop a much better understanding of the role of facilities in the business in order to grasp business continuity implications in the face of adverse events and plan accordingly (Pitt and Goyal 2004, Gill 2006). The challenges for facilities and business managers include the following:

- capacity to assess the risks associated with the impact of large-scale adverse events on business, people and facilities;
- planning and implementation of risk management measures in collaboration with relevant authorities and other partners;
- capacity to cope with business disruption, financial loss and the resilience to recover and prosper after the event;
- capacity to respond in a timely manner and implement business continuity measures;
- post-event financial and physical capacity to recover and restore business processes and facilities.

As noted above, the list of challenges is not exhaustive. However, it does serve to emphasize the point that facilities as a business resource have a number of dimensions, all of which have to be appropriately managed in order to ensure that facilities portfolios align with business requirements at economical cost and do not expose organizations to unnecessary risks. Facilities asset management provides a structured framework of managing facilities assets and enables organizations to meet the challenges outlined above.

3.3 Components of facilities asset management

As discussed briefly in Chapter 2, understanding the role of facilities in the business and the implications of facility decisions at various stages of the life cycle is fundamental to the effective management of facilities as a business resource. Hence the management approaches generally adopted for facilities asset management extend across the life cycle of the assets, divided broadly into performance review, planning, implementation and utilization phases. Each phase involves management and technical processes that result in specific outcomes and lead on to the next phase. We have adopted a similar approach. Figure 3.2 shows the components of the facilities asset management framework mentioned in Chapter 1. A brief overview of each aspect of facilities asset management is provided below.

3.3.1 Strategic facilities review

Strategic facilities review is the process of reviewing and assessing the performance of existing facilities against criteria that enable an organization to identify inefficiencies and opportunities for cost reductions as well as to develop an objective view of how well the facilities are meeting current business requirements. The review may also reveal their potential for meeting future needs. The results of

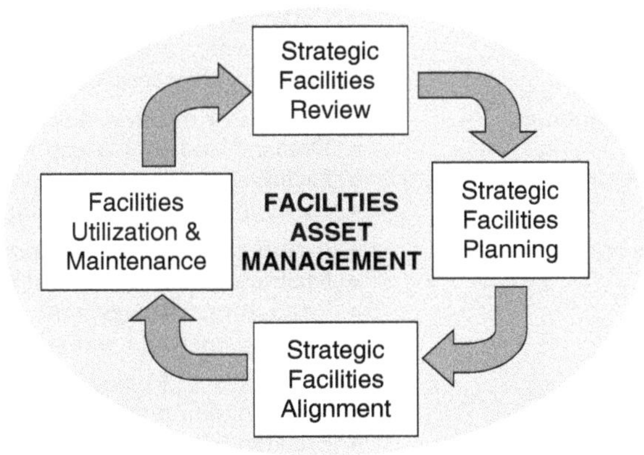

Figure 3.2 Components of facilities asset management

the review and assessment are contrasted against forecast business directions and objectives and thus form a critical input into strategic facilities planning.

3.3.2 Strategic facilities planning

This is the process of planning for facilities requirements across the portfolio of facilities owned or leased by an organization. It involves identifying strategic business directions and determining the facilities demand and supply situation with a view to identifying options for adjustment to the supply side of the equation to meet any changes in the demand for space. Demand management strategies may also be developed out of this process to keep demand within the boundaries of short-term capital constraints and long-term sustainability.

3.3.3 Strategic facilities alignment

Strategic facilities alignment is the process of implementing the agreed options from the strategic facilities planning process across the portfolio. The portfolio is fundamentally being realigned with the projected needs of the business through asset acquisition, leased space rationalization, asset renewal or disposal, or combinations of these actions. Strategic facilities alignment is a cyclical and iterative process that may extend over a number of financial years and planning periods.

3.3.4 Facilities utilization and asset maintenance

Facilities utilization and maintenance refers to the productive phase during which the facilities are occupied and used. During this (asset-in-use) phase the facilities

Table 3.1 Key elements of facilities asset management

Facilities asset management – components	*Key elements*
Strategic facilities review	• Facility portfolio analysis • Capability analysis • Performance improvement analysis • Scenario projections analysis
Strategic facilities planning	• Analysis of business direction and goals • Demand and supply gap analysis • Facilities options analysis • Development of facility alignment plan
Strategic facilities alignment	• Facilities portfolio rationalization • Facilities construction and refurbishment • Programme management • Facilities management support
Facilities utilization and maintenance	• Operation and management • Space planning and workplace management • Facilities asset maintenance • Facilities components renewal

need to be operated effectively and efficiently following facility operation processes that are legally compliant and in line with the organization's policies and procedures. Maintenance processes are also applied to maintain the assets in appropriate condition and enable the assets to be operated and used effectively. Effective management of space use, including demand and supply management, is also an essential aspect of this phase of the life cycle. Performance measurement of the facilities in meeting business requirements provides input into the strategic facilities review process and enables senior management to make facility decisions in a more informed and objective manner.

Table 3.1 is a summary of the key elements of strategic facilities review, strategic facilities planning, facilities portfolio alignment and facilities utilization and maintenance. The following four chapters will discuss these elements of facilities asset management in further detail. Suggestions for further reading are included at the end of each of these chapters.

References

Gill, T. J. 'Workplace Continuity: How Risk and Technology Will Affect Facilities Strategy'. *Journal of Facilities Management.* 2006. Vol. 4, No. 2, pp. 110–25.

Hinks, J. 'Business Strategy – Is FM a Core Concern?' *Facilities.* 2000. Vol. 18, Nos. 10/11/12, Guest editorial

Jensen, P. A. 'Organisation of Facilities Management in Relation to Core Business'. *Journal of Facilities Management.* 2011. Vol. 9, No. 2, pp. 78–95.

Jones, O. J. 'Corporate Partnership in Facilities Management'. *Facilities.* 1995. Vol. 13, No. 13, pp. 21–26.

Kaya, S., Heywood, C. A., Brawn, B. and Alexander, K. 'Raising Facilities Management's Profile in Organisations: Developing a World-class Framework'. *Journal of Facilities Management.* 2004. Vol. 3, No. 1, pp. 65–82.

Kincaid, D. 'Integrated Facility Management'. *Facilities.* 1994. Vol. 12, No. 8, pp. 20–23.

Krumm, P. J. M. M. and de Vries, J. 'Value Creation through the Management of Corporate Real Estate'. *Journal of Property Investment & Finance.* 2003. Vol. 21, No. 1, pp. 61–72.

Leaman, A. 'What is a Building For? Part I'. *Facilities.* 1991a. Vol. 9, No. 10, pp. 13–16.

——'What is a Building For? Part II'. *Facilities.* 1991b. Vol. 9, No. 11, pp. 16–19.

Loch, B. 'Avoiding the Usual Suspects'. *Facilities.* 2000. Vol. 18, Nos. 10/11/12, pp. 368–70.

Lynch, B. 'Maximising FM's Contribution to Shareholder Value Part 1: Can the Capital Expenditure Process for Fixed Assets Be Improved?' *Journal of Facilities Management.* 2001a. Vol. 1, No. 1, pp. 48–55.

——'Maximising FM's Contribution to Shareholder Value Part 2: Tactics for Improving the Capital Expenditure Process – and Producing Results'. *Journal of Facilities Management.* 2001b. Vol. 1, No. 3, pp. 222–36.

McGregor, W. and Then, D. S.-S. *Facilities Management and the Business of Space.* Arnold. 1999; Butterworth-Heinemann. 2001.

Nutt, B. 'Infrastructure Resources: Forging Alignments between Supply and Demand'. *Facilities.* 2004, Vol. 22, Nos. 13/14, pp. 335–43.

Pitt, M. and Goyal, S. 'Business Continuity Planning as a Facilities Management Tool'. *Facilities.* 2004. Vol. 22, Nos. 3/4, pp. 87–99.

Price, I., Matzdorf, F., Smith, L. and Agahi, H. 'The Impact of Facilities on Student Choice of University'. *Facilities.* 2003. Vol. 21, No. 10, pp. 212–22.

Roberts, P. 'Corporate Competence in FM: Current Problems and Issues'. *Facilities.* 2001. Vol. 19, Nos. 7/8, pp. 269–75.

Robertson, K. 'Work Transformation: Integrating People, Space and Technology'. *Facilities.* 2000. Vol. 19, Nos. 10/11/12, pp. 376–82.

Tay, L. 'Strategic Facilities Management of Suntec Singapore International Convention and Exhibition Centre – A Case Study'. *Facilities.* 2006. Vol. 24, Nos. 3/4, pp. 120–31.

Then, D. S.-S. 'The Role of Real Estate Assets in Supporting the Fulfillment of Corporate Business Plans – Key Organizational Variables for an Integrated Resource Management Framework'. *Facilities.* 2000. Vol. 18, Nos. 7/8, pp. 273–80.

Varcoe, B. 'Facilities Systems – What of the Future?' *Facilities.* 1993. Vol. 11, No. 10, pp. 7–11.

4 Strategic facilities review

As noted in Chapter 3, a strategic facilities review is the process of reviewing and assessing the performance of facilities against criteria that provide an informed perspective of how well the portfolio of facilities is meeting business requirements and at what cost. The assessment should also provide an indication of their potential in meeting future requirements. The main objectives of a strategic facilities review are to:

- evaluate the effectiveness of the facilities in supporting business processes, staff efficiency and productivity;
- assist in making decisions regarding the potential of the facilities to meet changing business directions and demand for space;
- identify any inefficiencies in the use of space and any opportunities for cost savings.

In essence, the strategic facilities review is a process of assessing current and future demand for facilities needs in line with projected business direction against the current supply of facilities. Its focus is on the capacity and capability of the current portfolio to meet current and future business facilities needs. The results of the review and assessment would feed into strategic facilities planning and thus form a complete loop for facilities asset management.

A strategic facilities review requires knowledge of the facilities at individual building level, facility level and portfolio level. It needs to be undertaken at portfolio level to avoid making decisions at facility level which may run counter to corporate strategies at portfolio level (for example, a business unit unknowingly disposing of a facility that may be required to be retained and reconfigured for a new business development as part of corporate strategy). Key elements of a strategic facilities review are outlined in Table 4.1.

4.1 Facilities portfolio analysis

The first step towards effective facilities asset management is to strive to attain comprehensive knowledge and understanding of the portfolio of facilities that the organization is using to support its business operations. A facilities portfolio analysis

Table 4.1 Key elements of strategic facilities review

Strategic facilities review – components	*Key elements*
Facilities portfolio analysis	Analysis of the facilities portfolio in terms of location, value, capacity, condition, utilization, operating costs and other attributes.
Capability analysis	Qualitative and quantitative evaluation of the capability of the facilities portfolio in meeting current business requirements.
Performance improvement analysis	Identifying areas for improvement in terms of utilization, functionality, reducing operating costs, energy consumption, etc.
Scenario projections analysis	Analysing the impact on the portfolio of various scenarios such as increase or decrease in space demand, fluctuation in the real estate market conditions, potential emergency situations and risks, etc.

is aimed at drawing together essential portfolio information that provides an inventory of current assets and lease commitments and the current operational profile of the portfolio. Typical information about the portfolio that would form the basis for a thorough facility portfolio analysis includes:

- *Portfolio footprint* (e.g. location, size, type, whether owned or leased, occupant)
- *Age* (e.g. chronological age, technological age, lease start/end dates)
- *Space* (e.g. land area, built-up area, net working space, leased space)
- *Production capacity* (e.g. process outputs, occupancy, estimated capacity – number of workstations, number of beds, etc.)
- *Capital value* (e.g. purchase cost, depreciated book value, market value)
- *Operating costs* (e.g. maintenance, utilities, security costs by month, year)
- *Quality* (e.g. quality grade, environmental star rating, market assessment)
- *Condition* (e.g. condition assessment rating, deferred maintenance backlog, capital replacements projections)
- *Functional suitability* (e.g. fitness for purpose, process efficiency, space quality)
- *Adaptability* (e.g. future development options – expansion, change of use, sale potential).

To enhance the analysis of the above information, the individual facilities making up the portfolio can also be classified into categories of criticality in contributing to fulfilling the organization's business purpose and objectives. This categorization would assist in the development of a more discriminative profile of the portfolio and allow the prioritization of improvement strategies and actions in accordance with facility criticality or importance.

4.2 Capability analysis

Based on the facility portfolio profile developed through a facility portfolio analysis, the organization is able to assess the strategic and operational capability and performance of the portfolio in supporting current business requirements. Evaluation criteria would include aspects such as:

- strategic positioning – distance to customers and suppliers, quality of supporting public infrastructure essential to the business, attractiveness to staff as a recruitment incentive and availability of quality staff;
- facilities performance – effectiveness in supporting and enabling business functions and processes (space capacity and quality, functionality, technology, design and layout, corporate image, etc.);
- financial performance – affordability of costs, cost comparisons with industry benchmarks, costs per unit of space or production, costs to restore condition, etc.;
- facility risks and liabilities – identification of risks and liabilities associated with physical condition and functionality such as deferred maintenance, overdue replacements, non-compliance with statutory requirements, environmental issues etc.;
- business alignment – degree of alignment of facilities resources with business directions and objectives (e.g. quality and quantity of space, technology, functionality, etc.).

A capability analysis provides the organization with a snapshot of the gap between the facilities required to support current operations with what is actually available.

4.3 Performance improvement analysis

A capability analysis would reveal to the organization areas of potential improvement and risks for which plans can be developed to address. Performance improvement analysis would usually focus on:

- improving efficiency by reducing the overall costs of facilities provision and operations – operating costs, cost of energy, facilities management services, etc.;
- initiatives aimed at improving utilization of space and reducing the cost of space (e.g. investments in technology infrastructure leading to changes to current space standards, rationalizing leases as a result of new workplace configuration);
- minimizing risk exposure associated with maintenance, repair and replacement issues, non-compliance with statutory requirements, indoor air quality, etc.;
- overcapitalization in assets resulting in capital tied down and unavailable for other more critical purposes or undercapitalization resulting in high leasing costs;
- reviews and modifications to improve overall organizational effectiveness in facilities resources utilization.

Actions to address the improvements identified can be segregated into priority categories of when they should be implemented, for example:

- as part of facilities management projects and further prioritized according to various criteria such as risk, criticality to business operations, benefits and costs, etc.; or
- brought forward into strategic facilities planning and addressed as part of a broad corporate strategy in aligning the facilities portfolio with new business directions and objectives; or
- addressed as and when opportunities arise (e.g. improved market conditions for sale of surplus property, consolidation of business operations, expiry of leases).

4.4 Scenario projections analysis

The business environment of most organizations is dynamic, and various scenarios will exist and change (predictably or otherwise) over time. The three preceding steps outlined above, facilities portfolio analysis, capability analysis and performance improvement analysis, would give the organization a sound grasp of its facilities portfolio and the ability to assess the likely implications of any changes in business direction. Scenario projections analysis is the process of identifying various scenarios (e.g. economic downturn, upsurge in services demand, changes in central bank interest rates, etc.) that might emerge and evaluating their potential impact on the business. Cascading this down to the impact on the facilities portfolio enables the organization to be better prepared for such scenarios should they occur and to plan for facilities strategies to respond to them.

This knowledge places the organization in a position where it can respond to any changes in a more measured and informed manner with greater chances of success in aligning its facilities portfolio to the new business directions and corporate objectives. It enables the organization to undertake strategic facilities planning effectively armed with accurate knowledge of its portfolio's strengths, weaknesses and adaptability to change.

4.5 Information needed for strategic facilities review

4.5.1 Portfolio information

Effective strategic facilities reviews rely heavily on facilities data and the collective knowledge of facilities management personnel, facility users, consultants and other stakeholders. The data needed is both quantitative and qualitative. The information required at the corporate level to undertake a strategic facilities review and make informed facilities decisions at portfolio level determines the types of data to be collected at building and facility level. These are usually drawn from the larger pool of financial and facility-related data collected and used at corporate and facilities levels for business planning, financial management and facilities operation. The important point to note is that usually most of the data will need to be drawn

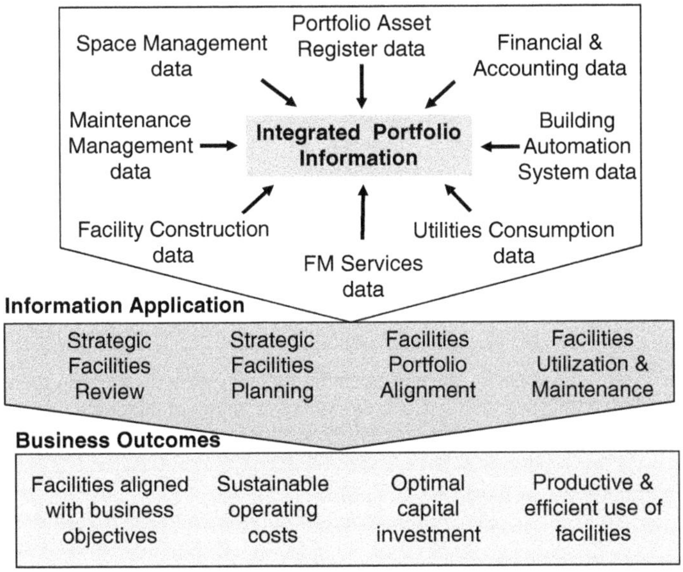

Figure 4.1 Integrated portfolio information – application and outcomes

from a number of transaction systems (such as the finance and accounting system, the space planning and management system, maintenance management systems, etc.) which very often are not integrated and hence the data would have to be extracted, integrated and combined with other knowledge to provide meaningful management information for informed decision-making and strategic evaluations. Figure 4.1 illustrates the relationship between integrated portfolio information and facilities asset management and the business outcomes arising from effective design and use of the knowledge base. The types of data shown are indicative of most portfolio situations.

The second key point to note is the need for careful design of an effective framework of monitoring and reporting on facilities assets that filters information from building level up to facility and portfolio levels. Much of the data is aggregated from individual buildings into a portfolio-wide database. The challenge is to define the data that will be used for standard and 'on demand' performance monitoring and reporting purposes at building, facility and portfolio levels to support strategic portfolio reviews, decision-making and strategic facilities planning.

4.5.2 Performance monitoring information

Typically, selected performance data at building level for major and significant buildings is aggregated upwards into facility-level monitoring and reporting. This enables comparisons to be made within the facility and decisions on individual buildings to be aligned with strategies at facility level (which in turn are driven by corporate facilities strategies). Similarly, selected performance data at facility level are summarized for reporting to senior management at the corporate level. This

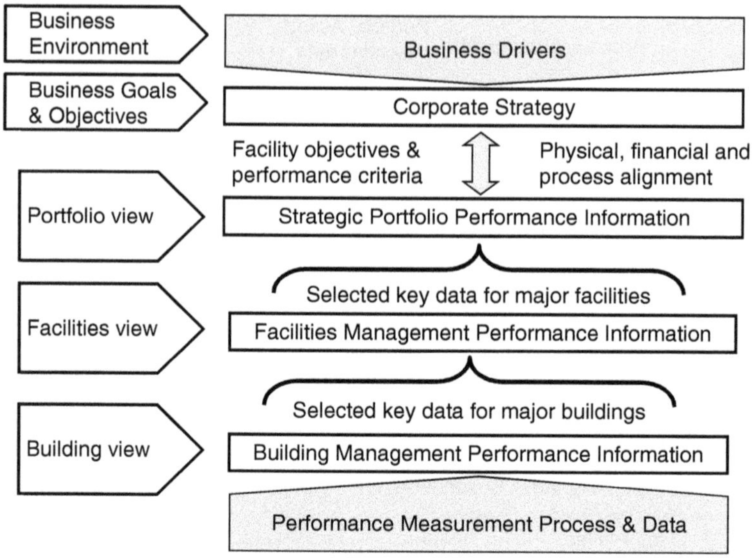

Figure 4.2 Hierarchy of facilities performance information

process ensures that the reports for various levels of management are not clogged up with unnecessary detail. The degree of aggregation and the types of performance data will depend on the organization and its performance management policies. Figure 4.2 illustrates the hierarchical flow of performance information to support monitoring and reporting, and in relation to requirements for strategic facilities reviews and strategic facilities planning.

4.6 Alignment with strategic facilities planning

The results of a strategic facilities review are a key input into strategic facilities planning, and therefore the planning and implementation of the review process should ideally be aligned with the needs of strategic facilities planning. This alignment is in terms of timing, scope and relevance of review results to planning needs.

Strategic facilities planning needs to fit in with the business-planning cycle, and timely information from a strategic facilities review is critical to ensure that strategic facilities planning is based on up-to-date and reliable information for demand and supply analysis. Hence the timing of strategic facility reviews must be carefully considered in the context of organizational planning in order to avoid unnecessary and costly repeats of data-gathering.

Useful reading for this chapter

Strategic facilities review

Apgar, M., IV. 'What Every Leader Should Know about Real Estate'. *Harvard Business Review*. Nov 2009, pp. 100–07.

Pertz, S. 'Redefining Strategic Facilities Planning'. *Facilities.* 1999. Vol. 13, No. 1, pp. 16–24.

Then, D. S.-S. 'The Role of Real Estate Assets in Supporting the Fulfillment of Corporate Business Plans: Key Organizational Variables for an Integrated Resource Management Framework'. *Facilities.* 2000. Vol. 18, Nos. 7/8, pp. 273–80.

Facility portfolio analysis

Becker, F. 'Integrated Portfolio Strategies for Dynamic Organizations'. *Facilities.* 2000. Vol. 18, Nos. 10/11/12, pp. 411–20.

Mitchell-Ketzes, S. 'Optimising Business Performance through Innovative Workplace Strategies'. *Journal of Facilities Management.* 2003. Vol. 2, No. 3, pp. 258–75.

Capability analysis

Rogers, P. A. 'Performance Matters: How the High Performance Business Unit Leverages Facilities Management Effectiveness'. *Journal of Facilities Management.* 2003. Vol. 2, No. 4, pp. 371–81.

Whitaker, M. J. 'Conducting a Facility Management Audit'. *Facilities.* 1995. Vol. 13, No. 6, pp. 6–12.

Performance improvement analysis

Amaratunga, D. and Baldry, D. 'A Conceptual Framework to Measure Facilities Management Performance'. *Property Management.* 2003. Vol. 21, No. 2, pp. 171–89.

McDougall, G., Kelly, J., Hinks, J. and Bititci, U. 'A Review of the Leading Performance Measurement Tools for Assessing Buildings'. *Journal of Facilities Management.* 2002. Vol. 1, No. 2, pp. 142–53.

Tan, T. H. and Then, D. S.-S. 'Assessing Building Performance – An Integrated Model'. Proceedings of 2004 CIB Congress, Toronto, Canada. 2–7 May 2004.

Varcoe, B. 'The Performance Measurement of Corporate Real Estate Portfolio Management'. *Journal of Facilities Management.* 2002. Vol. 1, No. 2, pp. 117–30.

Scenario projections analysis

Nutt, B. 'Infrastructure Resources: Forging Alignments between Supply and Demand'. *Facilities.* 2004. Vol. 22, Nos. 13/14, pp. 335–43.

5 Strategic facilities planning

Planning for its future facility needs is an important part of the business planning process of any organization intent on optimizing the contribution of facilities to its corporate objectives and goals. It is an integral component of business planning which also includes planning for the human resource, financial and information and communications technology needs of the organization. Effective planning enables the organization to ensure that it has the right configuration of facilities in terms of location, size, functionality, quality and other attributes to meet its business requirements efficiently and effectively. Strategic facilities planning is a planning process that anticipates and plans for the economical, orderly and timely adjustment of the facilities or portfolio of facilities to meet projected space, functional and other requirements. These requirements are the result of changes in business demand and direction in the context of prevailing and forecast business environments. Planning horizons can be from 2 to 10 years.

Strategic facilities planning occurs at portfolio level; that is, across a group of facilities that collectively provide the space and functional environment supporting the business. Facilities planning is also done at the individual facility level, based on the specific needs of the facility in response to local business conditions. However, this should not be done in isolation from the corporate strategic directions of the organization which could impact on the local provision for space. Planning at facility level is guided by the directions of the organization in the form of strategic business plans and other information. Conversely, strategic facilities planning must also be informed by facility plans developed at local level across the portfolio in order that any divergence between corporate and facility plans can be identified and adjustments made accordingly.

From the business management perspective, strategic facilities planning is a new way of thinking about the built environment and its relationship to organizational performance. It is a response to the growing awareness that facilities-related expenses represent most companies' second largest operating cost, next to staffing cost, and that facilities are their largest capital asset. It is the basis for effective decision-making by linking organizational goals and strategies to the design, development and utilization of their facilities as an enabling resource to achieve business outcomes. Figure 5.1 illustrates the context of strategic facilities planning within the overall business planning process.

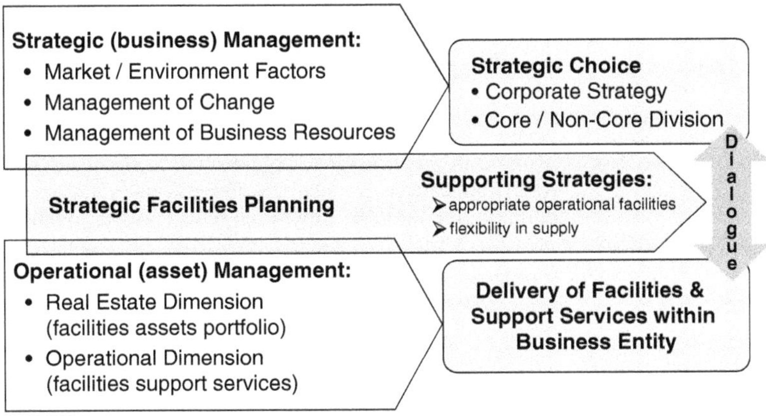

Figure 5.1 Context of strategic facilities planning

In practice, it is a framework for managing facilities in a way that supports an organization's long-range financial, operational and human resources goals. Strategic facilities planning involves three parallel processes:

- Monitoring how facilities are being used and managed.
- Evaluating whether facilities are best serving corporate objectives.
- Anticipating how facilities might better support the organization and respond to its changing needs.

Reasons for undertaking strategic facilities planning include:

- Facilities take time to plan, acquire or construct, have productive lives of many years and depend on effective planning based on correct projections of business demand to ensure they become available at the right time and continue to meet service requirements over their useful lives.
- Facilities are capital-intensive and large amounts of funds are usually not available at short notice. Therefore an effective strategic facilities planning process that informs the capital budget planning process is essential to ensure timely and adequate investment in facilities.
- The lack of or deficiencies in planning could result in over- or under-investment in facilities with business implications (loss of business opportunities, decline in competitiveness, failure to meet community expectations) and financial consequences (high capital and operating costs due to unused capacity).
- Strategic facilities planning enables organizations to proactively address emerging global issues such as environmental sustainability, financial risks, changing technology and increased threat levels to business continuity from factors such as severe climatic events and terrorism.

Detailed planning strategies and processes in individual organizations may vary due to factors such as business and portfolio size, financial capacity, nature of

Table 5.1 Key elements of strategic facilities planning

Strategic facilities planning – components	*Key elements*
Analysis of business direction and goals	Analysis of business direction and goals over the planning horizon in terms of growth, contraction or steady state and their implications on facilities dimensions (location, space demand, technology, functional requirements, quality, etc.).
Demand and supply gap analysis	Projecting future facility requirements using policies, standards, guidelines and other information. Comparing the projected needs with existing facilities to assess the gap between future demand and current supply.
Facilities options analysis	The development of facility solutions to close the gap between demand and supply, including analysis of financial and other implications.
Development of facility alignment plan	The development of capital works plans to deliver the approved facility solutions in step with projected business requirements. Includes development or adjustment of asset management plans.

business, organizational maturity and business growth or decline. However, the key aspects of the planning process applied in strategic facilities planning are fundamentally the same. Table 5.1 outlines the key elements of strategic facilities planning.

5.1 Analysis of business direction and goals

The business of an organization is shaped by its business environment, its strategies for responding to that environment and the business goals that it sets out to achieve. Business environments are dynamic and strategies to respond to changing environments have implications for facility assets in terms of demand for functional space. In addition these strategies can also involve factors such as incorporation of new technology to remain competitive, business process improvements, introduction of new or amended building regulations and responding to socially responsible issues such as environmental sustainability and climate change. All of these are facility-related and analysis of business directions is aimed at developing a comprehensive understanding of the demand for space as a result of the projected business directions. The implications on facilities may include:

- need for facilities in new locations;
- need for upgrading and improvement of existing facilities;
- need for new and more modern facilities (leased or owned);
- existing facilities becoming redundant; and
- potential for rationalizing spare capacity in the existing facilities portfolio.

Analysis of business direction and goals and their implications on facilities is complex and requires involvement of major stakeholders with intimate knowledge of business plans and strategies and the contribution of other stakeholders such as facilities users and facilities management personnel. The process varies from organization to organization but the aim is essentially identical – to understand the impact on facilities resulting from changes in business direction and objectives. It involves:

- analysis of strategic and operational business plans;
- discussions with key stakeholders to define and articulate business requirements;
- understanding of organizational policy in regards to issues such as environmental sustainability, climate change and risk;
- understanding of the functional and space relationships between business processes and facilities.
- understanding of strategic and operational constraints (internal and external).

A clear grasp of the implications of business direction on facilities enables a more detailed supply and demand analysis leading to potential facilities solutions to meet the projected business scenarios. Figure 5.2 outlines the concept of demand definition driven by an analysis of business direction and goals.

5.2 Demand and supply gap analysis

A core element of strategic facilities planning is the comparative analysis of business-driven demand for facilities and the capacity of the current portfolio of facilities (supply) to match that demand. The identified gap initiates the process of closing the gap and aligning the supply of facilities with the new level of demand. Effective facilities demand and supply analysis depends on having the necessary capacity (i.e. competences) to translate business processes and outputs (services or products) into facility dimensions of space, functionality, quality standards and other business-related facilities criteria. This translation process enables the anticipated

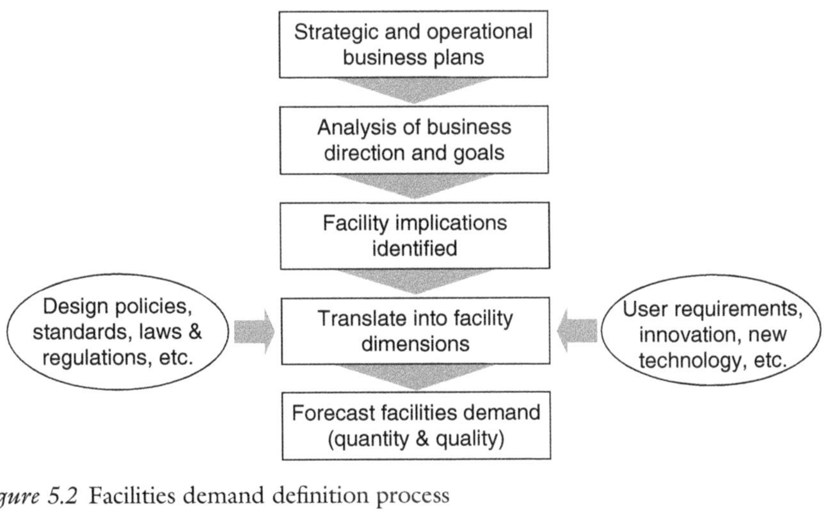

Figure 5.2 Facilities demand definition process

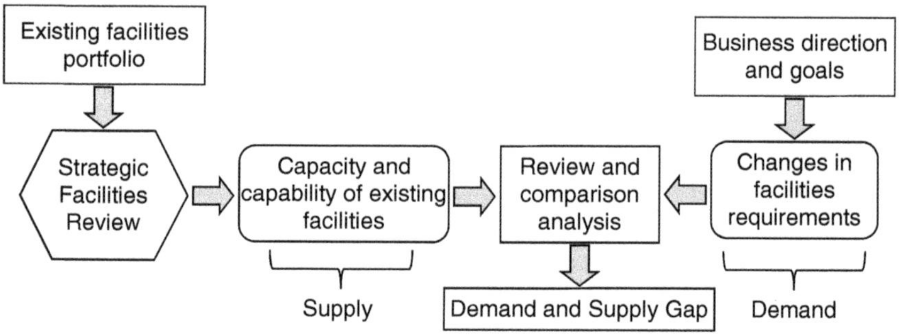

Figure 5.3 Demand and supply gap analysis

changes in facility requirements to be clearly defined, quantified and their estimated costs identified.

The second critical aspect of demand and supply gap analysis is to establish the capacity of the existing facilities to accommodate the projected facilities demand. The review may indicate the need for growth or, conversely, a reduction in facilities capacity. It may also indicate the need for only minor adjustments to the portfolio as a result of an anticipated steady state in business operations over the planning horizon. The result of assessing existing facilities against the projected facilities demand provides a snapshot of any mismatch between supply and demand. The information for the assessment is drawn from a strategic facilities review (see Chapter 4) as well as premises audits and space planning activities during the facilities utilization and maintenance phase which provides much of the information about the existing facilities. Figure 5.3 illustrates the process of demand and supply analysis based on strategic facilities review information and the implications of changing business direction and goals.

5.3 Facilities options analysis

The objective of facilities options analysis is to identify from the various options available the preferred facilities solutions to realign the existing facilities portfolio with future business directions and goals based on the gaps identified through the analysis of demand and supply. The gaps identified would not only indicate the shortfalls and surplus of space requirements but also the likely timing, constraints and opportunities for potential solutions. The options to address the gaps would normally include:

- acquisition of additional facility assets (new construction, purchase, new leasing);
- replacement of existing facility assets;
- disposal/demolition of surplus facility assets;
- leasing solutions (new, terminate, renew);
- refurbishment and modification of existing facility assets;
- elimination of the need for a facility asset solution (e.g. outsourcing a business process).

Options would be assessed using information on associated costs, time to deliver, benefits, risks and degree of alignment with business objectives. The key business issue is how robust the options analysis is and whether the analysis is focused on identifying the best fit with business requirements considering budgetary limitations and other constraints. A strategic facilities planning project team that operates as a collaborative effort between business managers, facility managers and other industry planning experts would provide the necessary expertise to evaluate the options and prepare recommendations and business cases for the preferred solutions.

A critical review of the solutions proposed would include the following questions:

- Have all possible options been identified?
- How robust was the evaluation process and what were the criteria and assumptions?
- Do the recommended options represent the optimal solutions to bridge the gap between facilities demand and supply?
- Has the evaluation process considered the implications of the facility plans of other business units in the group (for large corporations) or other government departments (public sector) on the recommended options and vice versa?
- Do the recommended options meet budgetary and other planning considerations?
- What are the overall capital and ongoing costs of adopting the recommended options?
- What are the anticipated implications of the financial market conditions if borrowing is required for implementation?
- What impact will the implementation have on business operations and will the timing of the implementation plan match the timing of business needs?
- What will be the implications of the real estate and property market conditions on the implementation of the proposed solutions?
- What impact will the proposed changes to the facilities portfolio have on existing facilities management capabilities and resources?
- What are the implications of the state of the construction industry in terms of competitive pricing and delivery capabilities?

5.4 Development of facilities alignment plans

The development of facilities alignment plans represents the final stage of the strategic facilities planning process. The facilities alignment plans should have at least two components:

- A capital projects plan – This is a plan that details all the capital projects that are proposed to be implemented as a result of the facility options analysis.
- An asset management plan – This plan articulates the strategy for managing, maintaining and operating the reconfigured portfolio and provides details of the changes to facilities management, financial and other resource provisions to support the changes in the facilities portfolio.

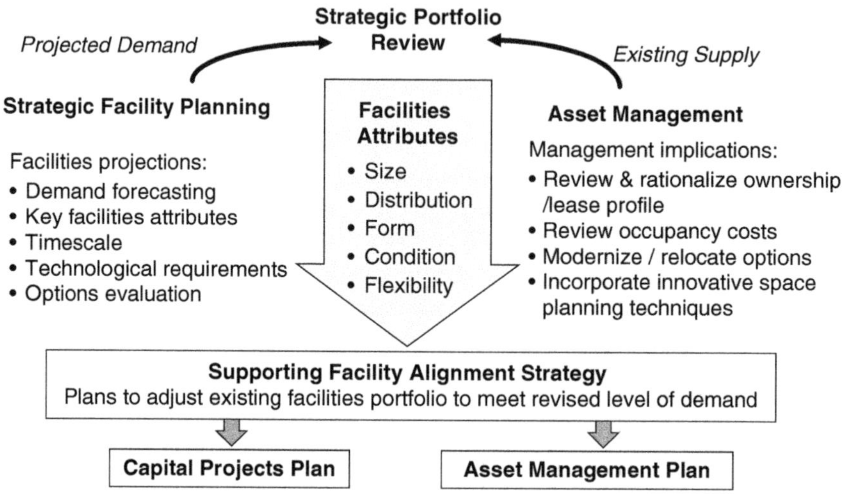

Figure 5.4 Development of facilities alignment plan

Figure 5.4 summarizes the inputs and outputs of development of the facilities alignment plan.

5.4.1 Capital projects plan

The most significant outcome of the strategic planning process is the capital projects plan which describes the organization's facilities alignment response to changes in business direction and goals. The capital projects plan may span several financial years and budgeting periods, requiring significant capital funds to implement and competent management for the delivery of successful projects. Due to the dynamic nature of the construction industry and rapidly changing technology, as well as the potential for more changes in business direction in the course of implementation, the plan needs to be flexible and capable of adaptation to changing circumstances (including cut-backs in funding).

The capital projects plan typically provides information such as:

- a list of the projects (facility, building, location, business division, etc.);
- details of the projects (new construction, acquisition, refurbishment, replacement, etc.);
- justification for each project (business case, cost–benefit analysis, etc.);
- project budgets, sources of capital, special financial conditions;
- commencement and completion time frames;
- implementation responsibility (project management);
- project delivery strategy (traditional project management, management contract, public–private partnership, etc.);
- any critical path considerations;

- project risks and mitigation strategies;
- the delivery responsibilities, management, monitoring, reporting and governance arrangements for the capital projects plan.

5.4.2 Asset management plan

Accounting practice typically separates capital funding from recurrent funding for facilities operation and maintenance. In practice, the capital projects plan is usually funded from sources of capital funds. On the other hand, maintenance and operation is funded by recurrent expenditure budgets which are often not appropriately adjusted to reflect any changes in the facilities portfolio even though the impact can be quite significant.

Leasing costs are also affected and budgets need to be suitably adjusted. However, while the funding of increased leasing costs is often not an issue due to legal contractual obligations, there is no such compelling incentive to allow for any projected increase in operating and maintenance costs. Facilities management departments often have to absorb these cost increases, adding further strain to budgets which could already be under pressure. Hence, formal recognition is necessary of the need for adjustments in maintenance and operation resources and budgets followed by action to make these adjustments. Plans for the realignment of the portfolio should therefore include not only the capital investment costs but also include budgetary and other resource adjustments to ensure that the reconfigured facilities portfolio can be sustained. An asset management plan is a mechanism to address this issue. The asset management plan should address matters such as:

- the changes in the maintenance and operations budget necessary to adequately cater for changes to overall facilities portfolio;
- the costs of collecting and updating facilities information, including any system requirements for accommodating new and additional information;
- any human resource and outsourced services implications for the facilities management department for providing support to capital works projects;
- proposals and budgets for addressing any life-cycle replacements and deferred maintenance projects;
- impact of the facilities portfolio reconfiguration on existing facilities, their site services and infrastructure and the costs of upgrading them to cope with increased demand;
- implications for existing contracts and leases and how these will be resolved; and
- upgrading of older facilities to meet new statutory requirements (usually triggered by any major refurbishments and modifications).

Useful reading for this chapter

Books

Barrett, P. and Baldry, D. *Facilities Management – Towards Best Practice.* 2nd Edition. Blackwell Publishing. 2003.

Edwards, V. and Ellison, L. *Corporate Property Management – Aligning Real Estate with Business Strategy.* Blackwell Publishing. Chapters 1–6. 2004.

McGregor, W. and Then, D. S.-S. *Facilities Management and the Business of Space,* Butterworth-Heinemann. Chapters 1, 4, 6 and 7. 2001.

Nutt, B. and McLennan, P. *Facility Management – Risks and Opportunities. Blackwell Science.* Part One: The Business Trail; Part Three: The Property Trail. 2000.

Rondeau, E. P., Brown, R. K. and Lapides, P. D. *Facility Management.* Wiley. Chapter 2. 2006.

Articles on strategic facilities planning

Glagola, J. R. 'An Introduction to Strategic Facilities Planning'. *Real Estate Issues.* Spring 2002. Vol. 27, Issue 1, pp. 13–15.

IFMA. *Strategic Facility Planning*: A white paper. 2009.

Klein, R. A. 'Strategic Facilities Planning: Keeping an Eye on the Long View'. *Journal of Facilities Management.* 2004. Vol. 2, No. 4, pp. 338–50.

Mount, S. K. 'Strategic Facilities Planning as Component of the Business Plan'. *Industrial Development.* August 1990. Document #50100 Conway Data Inc.

Pertz, S. 'Redefining Strategic Facilities Planning'. *Facilities.* 1999. Vol. 13, No. 1, pp. 16–24.

Springer, T. and Lockwood, S. 'Find the Way – Strategic Facility Planning'. *Today's Facility Manager.* Nov 2002. www.facilitycity.com/tfm/tfm_02_10_news2.asp (accessed 8 April 2005).

Summerell, R. 'Real Property Asset Management Planning'. www.businessfacilities.com/bf_05_07_news2.asp (accessed 16 February 2006).

Today's Facility Manager. 'The Impact of Strategic Facility Planning'. 2002. www.todaysfacilitymanager.com/tfm_02_12_topstory.asp (accessed 5 Oct 2005).

Analysis of business direction and goals

Hinks, J. 'Business Strategy – is FM a Core Concern?' *Facilities.* 2000. Vol. 18, No. 10/11/12. Guest editorial.

Price, I. 'Business Critical FM'. *Facilities.* 2004. Vol. 22, Nos. 13/14, pp. 353–58.

Articles on demand and supply gap analysis

Clark, E. and Hinxman, L. 'Developing a Framework of Competencies for Facilities Management'. *Facilities.* 1999. Vol. 17, Nos. 7/8, pp. 246–52.

Krumm, P. J. M. M., Dewulf, G., and de Jong, H. 'Managing Key Resources and Capabilities: Pinpointing the Added Value of Corporate Real Estate Management'. *Facilities.* 1998. Vol. 16, Nos. 12/13, pp. 372–79.

Nutt, B. 'Infrastructure Resources: Forging Alignments between Supply and Demand'. *Facilities.* 2004, Vol. 22, Nos. 13/14, pp. 335–43.

Roberts, P. 'Corporate Competence in FM: Current Problems and Issues'. *Facilities.* 2001. Vol. 19, Nos. 7/8, pp. 269–75.

Articles on facilities options analysis

Alexander, K. 'Identifying and Managing Facilities Needs'. *Facilities.* 1993. Vol. 11, No. 3, pp. 8–21.

Lehtonen, T. 'Collaborative Relationships in Facility Services'. *Leadership & Organization Development Journal.* 2006. Vol. 27, No. 6, pp. 449–64.

Varcoe, B. 'Business-driven Facilities Benchmarking'. *Facilities.* 1996. Vol. 14, Nos. 3/4, pp. 42–48.

Young, F. 'Integrated Operational Services: Meeting Continuously Changing Needs and Expectations'. *Facilities.* 2004. Vol. 22, Nos. 13/14, pp. 373–76.

Articles on development of facility alignment plan

Barrett, P. 'Achieving Strategic Facilities Management through Strong Relationships'. *Facilities.* 2000. Vol.18, No. 10/11/12, pp. 421–26.

Barton, R., Jones, D. and Gilbert, D. 'Strategic Asset Management Incorporating Ecologically Sustainable Development'. *Journal of Facilities Management.* 2001. Vol. 1, No. 1, pp. 74–80.

Becker, F. 'Integrated Portfolio Strategies for Dynamic Organizations'. *Facilities.* 2000. Vol. 18, Nos. 10/11/12, pp. 411–20.

Gaddie, S. 'Enterprise Programme Management: Connecting Strategic Planning to Project Delivery'. *Journal of Facilities Management.* 2006. Vol. 2, No. 2, pp. 177–89.

6 Strategic facilities alignment

Strategic facilities alignment refers to the process of making physical adjustments to the facilities portfolio so that it better aligns with the projected needs of the business. The need for alignment usually emerges from a strategic facilities review (e.g. underutilized space or lack of space, outdated technology, poor asset condition affecting business efficiency, etc.) and takes shape through strategic facilities planning (Chapter 5). Strategic facilities alignment involves planning and delivering a programme of capital projects made up of new facilities construction or acquisition, refurbishment, upgrades and expansion of existing facilities. Part of the programme could also include the disposal or demolition of existing surplus facilities to release capital which could be used for reinvestment in new facilities or upgrading of existing facilities. For organizations with facilities portfolios that include leased buildings, or relying solely on leased facilities for their business operations, the programme may include termination of existing leases and/or entering into new leases to realign space requirements with business operational needs. While strategic facilities alignment involves capital expenditure, there could also be opportunities for capital realization through asset sales and savings in space costs through rationalization of owned and leased space.

Strategic facilities alignment requires a broad range of real estate, building construction and facilities management expertise as well as programme and project management skills. These are supported by systems and processes related to financial management, facilities management, facilities design, construction, lease management and building information management. Table 6.1 shows the key elements of strategic facilities alignment.

6.1 Facilities portfolio rationalization

The outcome of the strategic facilities planning process presents the organization with a strategy for rationalizing its facilities assets (real estate) holdings across the facilities portfolio, extending over different business units or divisions at various

Table 6.1 Key elements of strategic facilities alignment

Strategic facilities alignment – components	*Key elements*
Facilities portfolio rationalization	Real estate activities relating to asset disposals, new acquisitions, land acquisition and sales, lease rationalization.
Facilities construction and refurbishment	Design and construction activities related to new construction, refurbishments, upgrades and expansion.
Programme management	Managing the entire programme of projects and budgets for strategic facilities alignment over the implementation period.
Facilities management support	Providing facilities management expertise to assist in real estate, design and construction projects, commissioning and handover.

locations in response to changing business directions and goals. These could include actions to:

- dispose of facilities that no longer meet operational needs or are surplus to requirements;
- acquire new facilities with more suitable functional attributes or in more strategic locations;
- sell off land that is surplus to requirements or no longer relevant to a strategic land holding and/or acquire other land (either for construction of new facilities or as part of a revised strategic land holding); and
- rationalize existing leased spaces and costs through lease expiry, terminations, extensions and new leases.

Depending on the scale of the facilities alignment program and the capacity of the department responsible for the rationalization process supplementation of in-house core skills with external expertise in real estate transactions, leasing and real estate marketing may be required. Disposals and acquisitions of facilities assets can evoke sensitive issues (particularly in the public sector) and part of the rationalization process would need to manage any implications arising from within the business units of the enterprise, the wider community or industrial relationships.

An accurate and reliable database of the current facilities assets portfolio is a prerequisite to the process of rationalization, and updating this database with details of the portfolio adjustments is one of the key functions of the facilities management department. Transaction data would also need to flow on to other systems such as the financial and accounting system as well as facilities management systems.

The rationalization process has significant capital and recurrent funding implications. These could be in the form of capital inflow from sales, capital outflows for acquisitions, commitments of recurrent expenditure for new leases or cost savings from termination of existing leases. Therefore a key part of facilities portfolio rationalization includes processes for managing the sources and application of capital funds and adjustments to recurrent expenditure budgets. For example, the proceeds from sales may need to be quarantined for supplementing funds for acquisitions. Savings from terminated leases in one business unit may need to be directed to other business units to pay for a new or expanded lease.

6.2 Facilities construction and refurbishment

Facilities construction and refurbishment projects usually form an integral part of strategic facilities alignment. The scale and scope of such projects depend on business circumstances and the outcomes of strategic facilities planning.

An essential competency required of any organization undertaking facility construction and modernization is the ability to prepare accurate project briefs to inform and direct design and construction professionals on project requirements. In terms of facilities dimensions, project briefs (or facility briefs) are projections of corporate image, driven by the organization's culture and policies and reflected in the use of and quality of space. These are articulated through its preferences in design and configuration of functional space; use of technology and environmental sustainability provision; and how the facilities assets are operated and managed as an ongoing process. Typically, the project briefs are prepared for scheduled projects in:

- refurbishments of existing facilities to modernize them, incorporate new functionalities and upgrade to new technology;
- expansion of existing facilities to provide new space capacity and functionality; and
- construction of new facilities.

Design and construction professionals rely on the project briefs to guide them in their work, and the quality of project briefs is critical to ensuring that the organization achieves the outcomes it requires. A key aspect of effective project briefs is clear policies and standards relating to space allocations (e.g. square metres per person), usage and functional definitions of how the space is to be fitted out and equipped.

A critical aspect of facilities construction and refurbishment is the procurement and management of the construction delivery process. Hence contract and project management are the other essential competencies for effective facilities construction and refurbishment. Procurement methods range from traditional construction contracts to alternative contracts which incorporate various features such as public–private partnerships and long-term maintenance or facilities management. The probity, transparency and accountability aspects of procurement are as important as the price, technical and contractual aspects. The intention here is not to delve into details about the contract procurement and management process but to highlight its importance and the need for expertise in this area for successful outcomes in facilities construction and refurbishment projects.

Advances in information technology have had significant impact on design and construction processes, resulting in increased efficiency and effectiveness in the way projects are delivered. From a facilities management perspective, the most significant impact is the amount and quality of information from the design and construction phases that is readily available for the utilization and maintenance phase – the longest phase of the facility's life cycle. Over the last two decades there have been tremendous advances in information technology applied in the design, construction and building management fields. The convergence of three-dimensional graphical capabilities and databases aided by the increasingly powerful and affordable processing capabilities of modern computers have led to the emergence of sophisticated asset and facilities management applications. Building Information Modelling (BIM) systems, which offer the capability to seamlessly capture data from the design stage through to construction and operation, could provide facilities managers with opportunities to manage from a position of knowledge of their facilities assets rather than being consumed by transaction management. However, the implementation of a deliberate and focused strategy to harvest design and construction information to support facilities management is subject to corporate management's acknowledgement of facilities assets as a business resource, as well as the ability and willingness of the design and construction professionals to share the information at affordable cost.

It is well established that the design and construction stage of a facility's life cycle presents the greatest opportunity to influence its construction and future life-cycle costs. Design and the selection of materials, plant and equipment influence the future costs of operation and maintenance, the economic lives of facility components and their replacement costs. Early involvement of building construction contractors in the design process helps to improve the buildability of the facility, and the input of facilities managers into design helps to improve the future maintainability, operational efficiency and functionality of the facility. The project design brief presents the opportunity to balance design innovation with a clear emphasis on future life-cycle costs considerations. With the growing awareness of environmental sustainability, the procurement of construction project delivery should not only be driven from a cost and value perspective, but should also take into account the environmental impact and the economy of future costs of operation, maintenance and replacement – i.e. a whole-life approach to facilities asset management is needed.

6.3 Programme management

Strategic facilities alignment is implemented through a programme of works relating to facilities asset rationalization, construction and refurbishment. The programme of works may span across several financial years and be delivered in different geographical locations, as well as for various business units and divisions of the organization. Such a diverse portfolio of projects requires effective management and coordination to ensure delivery on time, within cost budgets and in compliance with quality standards. Key elements of programme management include:

- programme planning based on factors such as priorities, availability of funds and critical path considerations;
- budget development, management and reporting to ensure that funds are available to pay for projects while minimizing the level of idle capital funds consistent with cash flow requirements and projects budget forecasts;
- implementation coordination, monitoring and reporting to track and report on progress of projects as well as managing the impact of any project issues, delays and cost overruns; and
- management of risks across the programme arising from factors such as contractor financial stress, industry skills and materials shortages, budget cut-backs, project reprioritization and industrial relations.

6.4 Facilities management support

Projects emanating from the strategic facilities alignment process are generally seen as the domain of the design and construction professionals. There is, however, a supporting role performed by the facilities management function which is important and necessary for the success of the strategic facilities alignment process. Facilities management inputs across a large programme of capital projects can be significant in terms of the diversion of facilities management resources from their designated functions. One aspect of the role of the facilities management function is their input into the design process regarding the future maintenance and operation of new facilities. However, the degree to which this role is embraced during the design process can be variable. Nevertheless, this is an important role and one that needs to be recognized, used and resourced. There are other aspects of this supporting role of the facilities management function which are equally important and together represent a significant additional workload which is often not compensated for in terms of either additional human resources or funding. The need to provide for the costs of temporary supplementation of resources should be recognized as part of the overall project cost and appropriately funded. Facilities management support of strategic facilities alignment capital work programmes includes the following:

- design reviews and inputs to improve facility functionality, efficiency, operation, maintenance and environmental sustainability;
- provision of relevant information on existing facilities assets including 'site intelligence' on services locations, etc.;
- coordinating site access, organizing planned services interruptions and liaison with facility occupants;
- logistical support in movement of occupants and furniture, etc.;
- coordinating connection of new services, including liaison with utility companies and local authorities;

- responding to facilities failures caused by accidental damage to pipes, cables, switchboards and other equipment in the course of the project;
- collecting textual and graphical facilities information during and after the project for facilities management purposes, including changes to facilities management systems to accommodate new data;
- witnessing testing and commissioning of the new and refurbished facilities;
- takeover of new and refurbished facilities;
- planning and establishment of resources, budgets and facilities management services for new and refurbished facilities;
- managing warranties and claims and monitoring of the rectification of construction defects;
- implementation of temporary measures to mitigate effects of any delays in the rectification of construction defects;
- decommissioning of facilities targeted for disposal;
- assisting in any post-occupancy evaluations of new and refurbished facilities.

Useful reading for this chapter

Books

Edwards, V. and Ellison, L. *Corporate Property Management: Aligning Real Estate with Business Strategy*. Blackwell Publishing. 2004. (ISBN: 0-632-06051-4).

Haynes, B. and Nunnington, N. *Corporate Real Estate Asset Management: Strategy and Implementation*. EG Books. 2010. (ISBN: 978-0-7282-0573-4).

Articles on facilities portfolio rationalization

Acoba, F. J. and Foster, S. P. 'Aligning Corporate Real Estate with Evolving Corporate Missions: Process-based Management Models'. *Journal of Corporate Real Estate*. 2003, Vol. 5, No. 2, pp. 143–64.

Apgar, M., IV and Bell, M. 'Managing Real Estate to Leverage Change: The Dun & Bradstreet Case'. *Site Selection*. Dec 1995, pp. 935–38.

Barrett, P. 'Achieving Strategic Facilities Management through Strong Relationships'. *Facilities*. 2000. Vol. 18, Nos. 10/11/12, pp. 421–26.

Barton, R., Jones, D. and Gilbert, D. 'Strategic Asset Management Incorporating Ecologically Sustainable Development'. *Journal of Facilities Management*. 2001. Vol. 1, No. 1, pp. 74–80.

Becker, F. 'Integrated Portfolio Strategies for Dynamic Organizations'. *Facilities*. 2000. Vol. 18, Nos. 10/11/12, pp. 411–20.

Gaddie, S. 'Enterprise Programme Management: Connecting Strategic Planning to Project Delivery'. *Journal of Facilities Management*. 2006. Vol. 2, No. 2, pp. 177–89.

Articles on facilities construction and refurbishment

Kashiwagi, D. and Byfield, R. E. 'Selecting the Best Contractor to Get Performance: On Time, on Budget, Meeting Quality Expectations'. *Journal of Facilities Management*. 2002. Vol. 1, No. 2, pp. 103–16.

Thomas-Mobley, L. and Khuncumchoo, N. 'A Facility Manager's Approach to Standardized Construction Contracts'. *Journal of Facilities Management.* 2006. Vol. 4, No. 4, pp. 234–44.

Articles on programme management

Bryde, D. J. and Joby, R. 'Incentivizing Project Performance in the Construction of New Facilities'. *Journal of Facilities Management.* 2007. Vol. 5, No. 2, pp. 143–49.

Greenwood, M. 'The Management of Project.' *Best Value in Construction.* Kelly, J., Morledge, R. and Wilkinson, S. Eds. Blackwell Publishing. 2002. Chapter 12, pp. 222–39.

Remington, K 'Project Management'. *Workplace Strategies and Facilities Management – Building in Value.* Butterworth-Heinemann. 2003. Part 2: Key Facilities Management Issues, Chapter 13 – Project Management, pp. 220–40.

Articles on facilities management support

Blyth, A. and Worthington, J. *Managing the Brief for Better Design.* Spon Press. 2001. Part One: Briefing Explained. Chapters 1–3, pp. 3–30.

Kelly, J. 'Project Briefing: Practice'. *Architect's Handbook of Construction Project Management.* Murray, M. and Langford, D. Eds. RIBA Enterprise. 2004. Part B – Inception and Briefing, pp. 43–66.

7 Facilities utilization and asset maintenance

A new facility commences its useful life phase once it is ready for occupancy and use. Similarly a refurbished facility has had its life extended and re-enters the utilization phase once more as soon as the upgrading work is completed. A purchased or leased facility may not be new physically and chronologically but transitions similarly to the utilization phase at acquisition. This is also the phase where facilities occupancy costs of operation and maintenance are continuously incurred as long as the facilities are in use as a supporting operational resource. These costs have been estimated to be approximately five times the initial capital cost. Therefore it is not difficult to understand the perception of operational facilities as a necessary cost burden with no tangible value-added and an asset with a depreciating capital value. However, there is an ongoing shift of perception from regarding operational facilities as a cost to business to that of a valued business resource. As described in Chapter 2, facilities assets, when competently managed, are increasingly seen as physical workspaces with the potential to improve human productivity by encouraging innovation and creativity as well as developing a competitive advantage through an enhanced corporate brand image in recruitment and retaining the best talent. For owner-occupiers, the long-term care of the assets in terms of facilities maintenance and judicious component renewals will not only preserve, but also enhance, the value of the assets by extending their economic lives. Figure 7.1 illustrates the critical twin roles of workplace management and asset management in facilities asset management over time.

Rapidly changing technology has transformed facilities from mere 'bricks and mortar' to technologically driven platforms that support sophisticated business processes and knowledge-based work. In parallel, concerns with environmental sustainability, climate change and escalating consumption of diminishing non-renewable resources have resulted in buildings which have been designed and built to standards governing levels of embedded energy and carbon footprints. Older portfolios are also under pressure to conform to new standards of environmental sustainability and performance, necessitating both component and technological upgrades to meet the demands of modern, flexible and innovative workplaces.

Physical space occupied by businesses, as a neglected resource, is perhaps more in evidence in recent years as a result of the global financial crisis which has presented many companies with surplus space as a consequence of strategies to reduce headcount, and such companies have come to realize the significance of facilities-related costs associated with occupied space. Successful implementation of

Figure 7.1 Context of facilities utilization and asset maintenance

appropriate workplace solutions will demand a thorough understanding of the drivers of business occupancy costs and how they can be systematically assessed, evaluated, procured and managed to remain strategically and operationally relevant in supporting the current corporate strategy. Success begins with strategic planning supported by integrated, robust processes.

In recent years, effective management of the utilization and maintenance phase of facilities assets has seen a general shift in focus from managing cost to managing performance. From this perspective, management of the facilities utilization and maintenance phase has the principal components and key elements shown in Table 7.1.

Table 7.1 Key elements of facilities utilization and maintenance

Facilities utilization and asset maintenance – components	*Key elements*
Operation and management	Operation and management of the facilities to enable occupancy and use.
Space planning and workplace management	Effective planning and efficient utilization of functional space within facilities to enhance productivity, innovation and creativity.
Facilities asset maintenance	Maintenance of the facilities to preserve their physical integrity and capacity to perform to standards relating to workplace quality, environmental sustainability and statutory compliance.
Facilities components renewal	Capital replacements and restoration as well as modifications to meet changing requirements and new legislation.

7.1 Operation and management

Occupants of facilities use the workspaces to conduct their business activities, and the objective of operation and management is to ensure that they are able to do so legally, safely and in an environment that is secure, comfortable, pleasant and functionally adequate to support their business processes. Operation and management is about managing over time with a focus on occupants' satisfaction with the facilities and services as an enabling workplace.

The basic elements of building services, such as air conditioning, lifts and escalators, fire protection, automatic doors, sound and video systems, security systems and lighting, need to operate reliably so that occupants can go about their business without disruption. Apart from compliance requirements of mandatory statutory codes and standards, there may be other standards imposed by production processes and/or by occupants to meet their business requirements (e.g. research laboratory indoor environment standards pertaining to indoor air quality, temperature, humidity, airflow and dust levels). Trained operators in control rooms aided by building automation systems are common in complex facilities.

Management of the facilities involves providing a range of services that support the occupants in their daily activities. These may include services such as:

- security and access control of the facilities (perimeter and individual buildings);
- reception desk services, including waiting areas and hospitality facilities for visitors and guests;
- help desk for maintenance, cleaning, catering, taxi, courtesy bus and other services;
- managing use of common spaces such as conference and meeting rooms and recreational areas;
- loading bay operations for material handling and logistics for deliveries of bulky items;
- mail room services for reception, security screening, distribution of incoming mail and dispatch of outgoing mail;
- conducting mandatory drills for fire evacuation and other emergency situations;
- services for special functions such as conferences and visits by high-profile persons;
- car park management and valet parking services.

Apart from the above, there is also a range of other 'behind the scenes' services that are provided and managed, such as waste management, cleaning, maintenance and testing of standby generators and alarm systems. Overall, the quality (adequacy, effectiveness and efficiency) of operation and management of the facilities determines how well the facilities support the business activities of the occupants. Occupancy also requires compliance with a wide range of legal requirements to ensure the health and safety of occupants and visitors. Many of these requirements are implemented as part of facilities asset maintenance (e.g. lift inspections and maintenance of fire protection equipment). An increasing number of buildings are also designed and built to environmental sustainability standards. Compliance with these standards relies on effective operation and management.

7.2 Space planning and workplace management

The provision of functional workspace is no longer simply determined by financial considerations. Demand assessment at the corporate level is multidimensional: it needs to take into consideration corporate image and culture, location preferences and transportation convenience, technological developments that support flexibility and mobility, and individual tasks and preferences. Clearly, facilities asset management is inextricably linked to the provision and management of business facilities in which the workplace and workspace environment represent the embodiment of the corporate culture and branding in terms of functional spaces.

There are numerous publications covering in detail the processes of space planning, workplace strategies and space management (Finch 2012, Meel, Nartens and Ree 2010, McGregor and Then 1999, O'Mara 1999, Vischer 1996, Becker and Steele 1995). This section aims to provide an overview of key elements of space planning and workplace management from the perspective of optimizing utilization of functional space within operational facilities assets as part of facilities utilization. Table 7.2 provides a summary of key elements in relation to space planning and workplace management. It should be noted that space planning and management during the utilization phase of the asset life extends into strategic facilities review by feeding relevant strategic data/information from premises audits and into strategic facilities planning by identification of accumulated forecast demand that cannot be fully satisfied through adjustments of existing spaces alone. For example, extensive maintenance repairs and or capital replacements may be required before the required spaces could become available, and such scenarios link space planning and management to other aspects of facilities asset management. These relationships need to be considered through a holistic approach so that space solutions that are developed deliver the optimum outcomes for the organization.

7.2.1 Strategic space planning

In today's business environment the business support infrastructure is no longer static. In a dynamic era when both planning and product cycles are shrinking, the element of time is becoming a major driver in terms of both space allocation and space utilization within the real estate portfolio of business corporations. Globalization, promoted by rapid technological advances in communication and connectivity, has intensified competition, adding to the ever-present demands for cost efficiency and higher productivity. In terms of physical asset infrastructure, corporate executives are increasingly forced to reassess their work environment: not only their location, size and internal configuration, but the fundamental economics of business resource utilization. The nature of work, way of working and work

Table 7.2 Key elements of space planning and workplace management

Space planning and workplace management	*Key elements*
Strategic space planning – the accommodation strategy	Understanding the status quo and projecting future requirements are key ingredients in effective space planning and management. A coherent corporate accommodation (facility) strategy is needed to drive decision-making.
Monitoring demand and supply of business space	Effective space planning necessitates a constant monitoring of demand and supply factors as inputs to the space planning processes. Demand is derived from organizational needs, while supply is a reflection of current facilities attributes. Successful space planning resolves supply and demand based upon an understanding of the present use of space, and in doing so establishes a profile for its future use.
Managing occupancy cost – monitoring utilization and managing space demand over time	The key to managing occupancy cost is monitoring utilization. Workplace management over time is about managing the dynamics of change from all dimensions – economic, technological, political and social.

environment are in a state of flux as organizations attempt to adapt to meet continuous change in global market trends and economies. Consequently, in order to ensure that the most fitting work environments are provided to meet the needs of business units, corporate executives (and/or their facilities managers) tasked with the responsibility of developing workplace strategies and planning workspace must be constantly considering their potential changing needs for the future. The workplaces of the future must not only accommodate rapid changes – of a political, economic, technological and social nature – but must also strive to reflect and promote new ways of working. The design and provision of a correctly balanced infrastructure of physical and virtual workspaces will be at the heart of strategic space planning for modern corporations. The importance of having a coherent corporate accommodation (facility) strategy that drives decision-making is paramount in managing space demand over time. The accommodation strategy will drive the provision of appropriate workspaces as an integral part of their overall business delivery strategy. For facilities management, although the traditional pressure has always been to reduce costs, decisions are now being taken in the light of maximizing value and competitive advantage. The modern role of strategic space planning embraces all matters of infrastructure and supports business by enabling effective business practice through the provision of appropriate facilities. It embraces issues of organizational reform, worker productivity, information technology, sustainability and occupational health and safety. Hence, facilities and the workspaces within them need to be designed to be flexible enough to be able to adapt to changes in the above issues without unnecessary waste and cost.

The design and reconfiguration of space within organizations, whether in terms of relocation or space reconfiguration, involves not only physical changes but also emotional changes for the people affected. Effective facilities change management will require a thorough process of assessing organizational needs, assessing supply and deriving an appropriate outcome through matching demand and supply. Success in such a process will rely on taking a vision of the future and a structured approach, supported by a comprehensive understanding of the business needs, data from the organization and physical characteristics of the building infrastructure (including constraints). In space planning and management the focus must be on the people using the facilities, the technology needed to support tasks and work patterns and the management skills, together with the necessary systems capabilities to adjust and monitor facilities performance.

One of the key responsibilities of facilities managers in managing demand over time is to ensure that the original assumptions that formed the basis of the accommodation strategy are still valid, and remain so when set against the ever-changing business environment. All too often, facilities management is reduced to maintenance, engineering or office services, but space management is a strategic role. The provision of workspace is a strategic issue because of the need to manage change that can occur at three levels:

- changes to the business – its operations, expansion or contraction of the business and its constituent parts, and the identification of business needs as they vary through time;
- changes to work practices – the adoption of new work methods and processes, the introduction of technology to automate or obviate manual activities, and the changing expectations of the people in the business;
- changes to workspace – the process of reconfiguring existing workspace, relocation to new premises and changes to the way workspace is allocated.

The process of maintaining the strategic relevance of the accommodation strategy is a continuous one. Any review of an accommodation strategy today must consider the implications of the above factors on the existing operational portfolio and the likely future need for other buildings. While such a review is an integral part of the process of strategic facilities planning, effective utilization of the facilities portfolio requires ongoing vigilance on the effectiveness of the accommodation strategy.

7.2.2 Monitoring demand and supply

The planning of all workspace should be a response to specific business needs and corporate drivers – i.e. organizational needs. The challenge of managing workspace is to achieve the best match of the supply of available functional space to the dynamics of changing business demand. The underlying business rationale of effective workplace management is the promotion of business effectiveness.

The effective management of workspace is based upon two imperatives – first, the acceptance of the economic reality that every square metre of occupied space has to be paid for by the business; and second, that the space occupied by staff should be directly related to the requirements of the work processes and tasks the person has to perform, rather than based primarily on their seniority and status. It is from this starting point that the relationship between business planning and space planning becomes obviously clear.

The process of addressing the workspace needs of the business is the conscious planning and design of the work environment to facilitate the delivery of products and services, driven by desired business practices (culture and organizational structure) and operational requirements (tasks and functions). This process is data-intensive and is structured to gather information about the current and future needs of the business, and the users within it. The level of detailed information required will vary depending upon the stage of the planning process, i.e. strategic or tactical. Typically the types of data/information that the facilities manager will require to enable planning of workspaces to be carried out will be grouped under people, workspace and services. The data groups are assessed at work group level, and by aggregation of the needs of each group, including at the organization level. The outcome of demand assessment is a statement of the workspace requirements for the organization as a whole, the synthesis of which is used to develop an accommodation plan. In gathering the data, whether at strategic or tactical levels, it is of vital importance to distinguish between the 'wants' and the 'needs' of the business. The use of appropriate data-gathering tools can go a long way in assisting the facilities manager with the screening out of undesirable 'wish lists'. Figure 7.2 illustrates key elements of workplace design data.

While the translation of organizational needs is reflected in elements of workplace demand (Figure 7.2), the premises audit is the supply side of the space planning equation. The supply variables are concerned with factors that influence decisions about the provision of real estate resource and the servicing of workspace in support of business processes. The objective of a premises audit is to assess how well a building is currently meeting the needs of the occupying business, and the degree to which it could satisfy their future needs. Many of the issues are similar to those that would require to be undertaken as part of the process of evaluating new premises. The key issues of a premises audit will encompass location, condition, utilization and value for money.

Effective utilization of the workspaces within premises is determined by the suitability of the environment for the work tasks to be performed. Therefore in constructing procedures for the conduct of premises audits, it is essential that the needs of the business are known as they will act as the backdrop against which the evaluation process will be conducted. Consequently, the premises audit will, in addition to assessing the condition of the fabric and services of the building, also take account of how the premises are used. Analysis of the audit data will enable the facilities manager to identify aspects requiring attention, such as the maintenance of the building's fabric, improvement of services and the optimization of workspace layouts. The single most important aspect of the premises audit is to

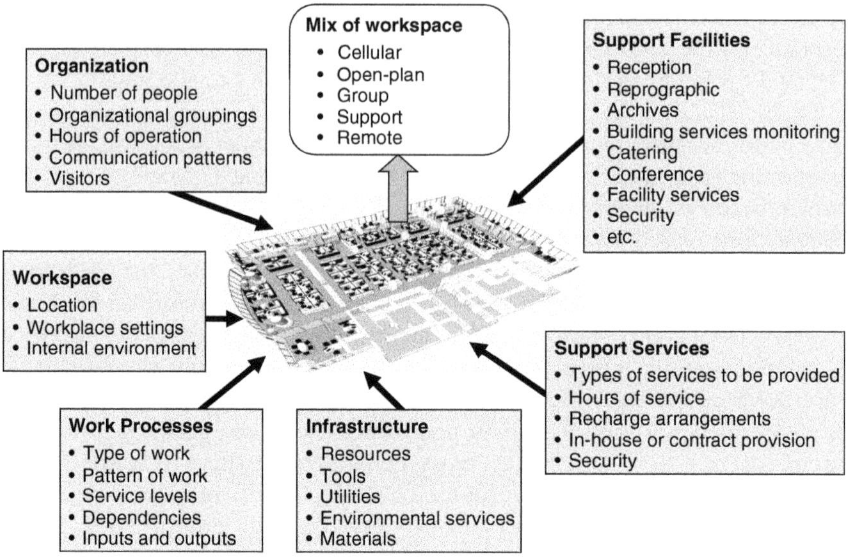

Figure 7.2 Elements of workplace design data (Then 2012, summarized from McGregor and Then 2001, Chapter 4)

identify what requires to be addressed to ensure that the occupier is receiving the best value for money that they can from the premises. This will include assessments of energy efficiency, adaptability and cost in use, bringing together all the strands of the audit, i.e. location, condition and utilization. The audit should therefore include an assessment of the costs of implementing workplace changes, be they partition moves, data/communications cabling reconfigurations or others, and assess them against the provisions made for flexibility, e.g. the provision of, and investment in, relocatable partitions, raised floors and the like. To summarize, in assessing the supply side of space planning, the functionality and quality of facilities are at least as important as the amount of space to be occupied.

Successful space planning resolves supply and demand based upon an understanding of the present use of space and in doing so establishes a profile for future use. As a strategic asset, as well as an operational resource, the provision of appropriate workspaces and the planning for their change and adaptation over time are critical if disruption to the business is to be avoided. Perhaps, even more significant, is senior management's endorsement that the working environment can be used as a catalyst for change. The challenge is to effectively match supply to demand for space over time guided by an agreed strategic framework and an objective demand-led analysis of workspace requirements.

7.2.3 Managing facilities occupancy costs

All businesses require real estate or facilities to operate from. Yet the fact that it is one of the most undermanaged corporate assets has not changed much since the early 1980s when the issue was highlighted by two studies from Harvard University (Zeckhauser and Silverman 1981) and the Massachusetts Institute of Technology (Veale 1989).

In a recent article in the *Harvard Business Review* (Nov 2009), it was stated that:

> For most businesses, it [real estate] is the largest or second-largest asset on the books, yet, because it is everywhere, real estate is easy to take for granted … . Business real estate is not merely an operating necessity; it's a strategic resource. But it rarely captures senior management's attention.
>
> (Apgar 2009, p. 100)

Based on the above publications and the quote almost three decades later, there appeared to be little change in the fact that corporate real estate assets are generally undermanaged and, as a result, sub-optimization in the utilization of the real estate resource still appeared to be prevalent in many organizations. However, the pressure to reduce operating costs has, in recent years, shifted from the initial focus on reducing overall staff numbers (headcount reduction) to initiatives aimed at reducing premises occupancy costs. The pressure to reduce occupancy costs has resulted in the scrutiny of two main aspects – the amount of space occupied by business units and the utilization of existing spaces.

Utilization is about achieving the optimum match between timing of need and availability of supply. In terms of space planning and management, the objective is to aim for the best match between demand for workspace to support business activities and its availability in terms of timing and duration of requirements. Demand for workspace relates to appropriateness of the types of space and their distribution within a building or a portfolio of buildings. Supply relates to the pattern of use of the available space and the range of tasks it is supporting.

No matter how well a workplace is engineered to meet the needs of its user, it adds no value to the business when the space is not in use. Experience from many office workplace utilization surveys shows that, for activities which are 'desk bound', occupants rarely spend more than 80 per cent of their time actually at their workplace, i.e. a desk. In practice, it is not uncommon to encounter many businesses where the utilization of their workplaces falls between 40 per cent and 50 per cent. Such situations have led more and more businesses to adopt 'flexible' styles of working or 'alternative workplaces', where the focus is on providing a variety of works settings from which the user is free to choose. (For examples of alternative workplace implementations please refer to the excellent *Harvard Business Review* paper by Apgar 1998.)

Over the last decade or so, the pace of development in information and communication technologies has dramatically impacted on the way people live and work. Particularly, in the commercial office sector occupied by 'knowledge workers' the space utilization patterns are changing with potentially dramatic implications for the way business enterprises will procure and occupy their office buildings. The drivers underpinning this change in space utilization pattern are a combination of economic pressures, generational changes in the workforce, organizational transformation and a growing environmental/sustainability agenda. Technology is directly or indirectly acting as a key enabler for all the above drivers, the main ones of which are:

- economic – as organizations respond to competitive pressures to become more cost effective, attract talent and be more agile;
- organizational – as they transform their structure to be flatter, and work processes to embrace greater flexibility and mobility in order to respond to a fluid business environment;
- demographic – a result of the changing demographic of workforce and the need to cater for greater diversity in ethnicity and inter-generational mix. For example, the Generation X workforce comes with a whole new attitude to work, having had a taste of flexible working and now demanding it.

Utilization is not only a function of efficiency of layout, but is also derived from the degree of flexibility the premises afford the users in terms of their capability to support the relocation of working groups, departments and functions (i.e. churn), and also to meet changing patterns and practices of work, including remote working. For many corporations, large expanses of sterile, production line-style office space (largely under-occupied) are yielding to a more dynamic work environment in which teamwork, collaboration and meeting space occupies a far greater proportion of the total space. The fundamental shift in attitude is changing from allocating workplace on a dedicated basis to a more flexible approach where more people can be supported by the same number of workplaces. Indeed, for many modern corporations, the workplace has taken on a new geography characterized by being location-independent and user-driven, fully functional and supportive, and occupied on an intermittent pattern and for varying durations. These new 'open and shared' configurations of corporate spaces represent a new dynamic in the traditional understanding of occupancy densities as they are not only occupied more extensively, but they are also utilized more intensively. There are major implications arising from these changes for those involved in the future planning and provision of office space. Not the least of these is the growing importance of workplace planning, in which resource managers from real estate/facilities, technology and human resources must work in concert to provide a much more dynamic workplace proposition, with implications as to how best to measure space utilization compared to the current status quo.

Given the pace of technological development in information technology, which is relentless, and will continue to affect people's lives and how they work, it is clear that changes in the workforce and the nature of work are inevitable. The key variable is how businesses will adapt to the changes, and the pace of change they choose to adopt. For real estate and facilities management professionals, one of the main challenges is to develop the competencies that are necessary to facilitate the business cases for transforming current practices and managing the changes to their workplace portfolio and facilities services that will result in improving the overall corporate performance.

7.3 Facilities asset maintenance

There is ample literature on asset maintenance management that describes the detailed processes and strategies that can be applied to facilities and the various asset components (Wordsworth 2001, Chanter and Swallow 2001). Therefore it is not the intention here to provide more of the same. However, a systematic approach towards an effective asset maintenance regime is discussed in the context of a facilities asset as a business resource. An effective regime of asset maintenance provides assurance to the organization that its facilities would continue to serve its business requirements, meet compliance requirements and perform to the required standards of contemporary workplaces as well as environmental sustainability criteria.

From a business perspective, facilities asset maintenance plays a key role in the following aspects of facilities ownership and utilization:

- business continuity and workplace availability – ensuring that the facilities are fit to be occupied and used in accordance with their intended purpose and minimizing interruptions to business activities caused by building-related failures;
- preservation of asset condition and value – preserving the condition (and hence value) of the assets through timely and effective repairs and replacements as well as regular servicing, testing and inspections;
- statutory compliance – compliance with various legal regulatory requirements and codes relevant to building occupancy and operation;
- operating costs – minimizing costs of premature replacements and repairs as well as improving the efficiency of energy and water consumption;
- facilities image and branding – maintaining the quality and presentation of facilities and workplace environment to designated standards of appearance, condition and ambience.

Four components that are important in establishing an effective asset maintenance regime are discussed in the following sub-sections. These are summarized in Table 7.3.

7.3.1 Asset maintenance policy

An asset maintenance policy articulates the organization's attitude towards the maintenance of its facilities assets and sets the direction for maintenance actions by expressing the desired outcomes of asset maintenance expressed through standards of asset condition, space allocation and service quality for the types of facilities in the portfolio. The asset maintenance policy should take into consideration policies on other facility-related matters such as design, procurement and construction of

Table 7.3 Key elements of facilities asset maintenance

Facilities asset maintenance – components	*Key elements*
Asset maintenance policy	A maintenance policy acknowledges the role of maintenance and expresses the organization's commitment towards the maintenance of its facilities.
Asset maintenance framework	A formal structure of strategies and processes that guide maintenance planning, implementation and management.
Maintenance capability	The capability of the organization to undertake maintenance, including information systems, budgets, workforce and equipment.
Maintenance implementation	The management, delivery and performance aspects of maintenance implementation.

facilities assets which have significant influence on future maintenance, facilities longevity and operational performance. A key aspect of the asset maintenance policy is that it determines the framework for the planning, implementation and management of maintenance. A well-articulated asset maintenance policy sends a clear message that maintenance is important to the organization in terms of business-focused outcomes and cost efficiency and encourages innovative solutions to improve both. The absence of a policy leaves the organization exposed to the possibility of a lack of focus in maintenance activities.

7.3.2 Asset maintenance framework

An asset maintenance framework defines the maintenance strategies to be adopted in accordance with the asset maintenance policy and the maintenance planning, implementation and review processes required to support and execute the strategies. It provides for consistency across the portfolio and enhances the ability to make effective maintenance-related decisions by using reliable information. The framework should describe in broad terms the following:

- maintenance objectives and outcomes as articulated in the asset maintenance policy;
- maintenance strategies in relation to the appropriate balance between categories of maintenance actions (i.e. planned, preventative, reactive and statutory compliance);
- methods for classification of asset condition and identification of maintenance work demand (e.g. condition standards, condition surveys, plant monitoring, etc.);
- a maintenance prioritization method based on condition, risk, asset criticality and other criteria;
- maintenance performance criteria and minimum achievement benchmarks;

- maintenance implementation strategies;
- maintenance review and reporting requirements.

7.3.3 Maintenance capability

An asset maintenance policy and framework for implementation would be ineffective without the appropriate level of organizational capability to implement the policy and deliver the work required. Maintenance capability includes adequate budgets, maintenance management and information systems, equipment, human resources with the necessary skills and competencies, and a reliable, competitive supply chain of equipment manufacturers and suppliers, contractors and specialist consultants. Capability also includes the way in which these resources are organized and deployed for maintenance purposes. However, this will depend on the particular maintenance management model adopted. The capability would be built up based on an assessment of the nature, size and condition of the asset portfolio and the policy requirements.

7.3.4 Asset maintenance implementation

Asset maintenance implementation refers to the management functions, procurement, delivery and performance aspects associated with facilities asset maintenance. The procurement model adopted (in-house, outsourced or mixed mode of delivery) determines the management role and functions necessary. This could range from a largely contract management role (in a totally outsourced model) to one that is fairly extensively involved in the implementation process (mixed model). The asset maintenance policy and other organizational policies would determine the nature and level of outsourcing to be applied. The actual delivery of planned and unplanned maintenance, repairs, condition surveys, testing, inspections and servicing is dependent on the way in which the workforce is organized, deployed and coordinated as well as the nature of the portfolio, its geographical dispersion and other logistical considerations. The key point to note is the degree of integration of the organization of the implementation effort with policy within a structured framework. The effectiveness of the integrated whole is measured through a performance measurement system.

An effective system of performance measurement provides the organization with feedback on whether maintenance outcomes as defined by the asset maintenance policy have been achieved, where shortcomings may have occurred and why. A performance measurement system allows the organization to review its asset maintenance efforts with a view to improvement and adjustment where necessary. It may also be used as the basis of a performance-based maintenance contract in which the client specifies the required performance levels and the contractor determines the processes necessary to achieve the agreed performance standards. Innovative results and radical new solutions can be achieved by taking a performance-based approach within an appropriate facilities management framework. Figure 7.3 summarizes the key components of effective asset maintenance.

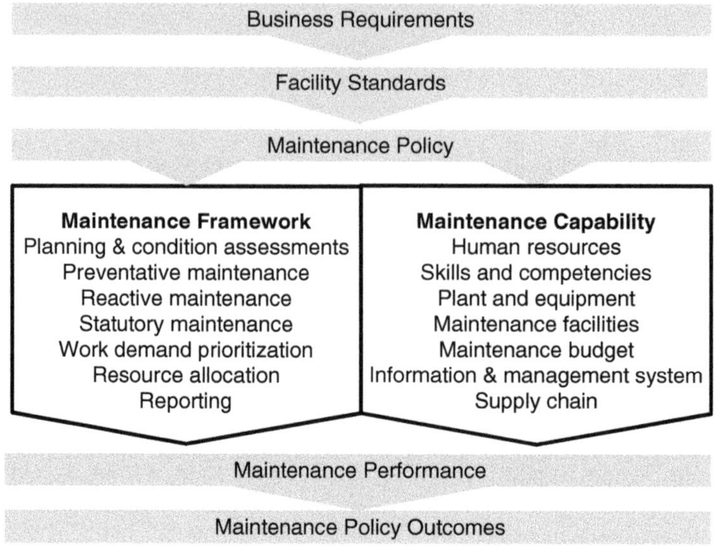

Figure 7.3 Facilities maintenance components

7.4 Capital replacement planning

Facility assets are comprised of component systems such as lifts and heating, ventilation and air conditioning (HVAC) systems which have physical and technological lives that are quite different from the life of the overall structure itself. Many of these will degrade in performance, repeatedly fail or become technologically obsolete during the operational life of the facility. The replacement and upgrading costs of the major component systems such as power and air conditioning systems, roofs and access control systems incur periodic cash injections which are usually not factored into maintenance budgets unless they have been identified and planned for. Accounting treatments of such cash injections typically classify them as capital rather than recurring expenses. Capital costs are generally excluded from maintenance budgets (expense) and have to be bid for against other competing priorities. As a result, capital replacements tend to be *ad hoc* and reactive. The outcome of such a reactive response to the replacement of key component systems is likely to be facilities which are underperforming in terms of condition, appearance, reliability of services and meeting health and safety standards.

Capital replacement planning is proactive asset management and a conscious effort at identifying the need for major replacements and upgrades with a view to forecasting and securing capital budget commitments for implementation before asset performance reaches unacceptable levels. Based on the planning outcomes, programmes for timely replacements can be planned and carried out. Capital replacement costs, together with the annual cost of maintenance, are also key inputs into maintenance plans and decisions about the future of the asset in the context of strategic facilities planning.

In brief, capital replacement planning consists of:

- identifying major components and systems requiring replacement during the useful life of the asset;
- establishing their design-useful lives from industry and technical data;
- preparing a plan for their replacements based on an assessment of their remaining useful lives, condition, workload, operating environment and other factors;
- preparing time-adjusted cost estimates for replacement at the planned dates;
- developing multi-year budget plans for the identified replacement costs;
- incorporating approved replacements into maintenance plans and capital project plans within the overall strategic facilities plan;
- implementing the approved capital replacement programme, including a pre-implementation review process to confirm replacement timing and priorities.

References

Apgar, M., IV. 'The Alternative Workplace: Changing Where and How People Work'. *Harvard Business Review*. May–June 1998. Vol. 76, No. 3, pp. 121–36.

—— 'What Every Leader Should Know about Real Estate'. *Harvard Business Review*. Nov 2009, pp. 100–07.

Becker, F. and Steele, F. *Workplace by Design – Mapping the High-Performance Workscape*. Jossey-Bass. 1995.

Chanter, B. and Swallow, P. *Building Maintenance Management*. Blackwell Science. 2001.

Finch, E. Ed. *Facilities Change Management*. Wiley-Blackwell. 2012.

McGregor, W. and Then, D. S.-S. *Facilities Management and the Business of Space*. Arnold. 1999. Reprinted under Butterworth-Heinemann. 2001. (ISBN: 0-340-71964-8).

Meel, J., Martens, Y. and Ree, H. J. *Planning Office Spaces – A Practical Guide for Managers and Designers*. Centre for Facilities Management, Technical University of Denmark. 2010.

O'Mara, M. A. *Strategy and Place – Managing Corporate Real Estate and Facilities for Competitive Advantage*. The Free Press. 1999.

Then, D. S.-S. 'The Business of Space'. *Facilities Change Management*. Finch, E. Ed. Wiley-Blackwell. 2012, pp. 57–75. (ISBN: 978-1-4051-5346-1).

Veale, P. R. 'Managing Corporate Real Estate Assets: Current Executive Attitudes and Prospects for an Emergent Management Discipline'. *Journal of Real Estate Research*. 1989. Vol. 4, No. 3, pp. 1–22.

Vischer, J. C. *Workspace Strategies – Environment as a Tool for Work*. Chapman & Hall. 1996.

Wordsworth, P. *Lee's Building Maintenance Management*. 4th Edition. John Wiley & Sons. 2001.

Zeckhauser, S. and Silverman, R. 'Rediscover Your Company's Real Estate'. *Harvard Business Review*. Jan–Feb 1981, pp. 111–17.

Useful reading for this chapter

Books on space planning and workplace management

Grantham, C. *The Future of Work*. New York. McGraw-Hill. 2000.

Malone, T. W. *The Future of Work*. Boston. Harvard Business School Press. 2004.

McGregor, W. and Then, D. S.-S. *Facilities Management and the Business of Space*. Arnold. 1999. Reprinted under Butterworth-Heinemann. 2001. (ISBN: 0-340-71964-8).

Books on facilities asset maintenance

Chanter, B. and Swallow, P. *Building Maintenance Management*. Blackwell Science. 2001.

Then, D. S.-S. *Real Estate Asset Management – Operational Property Assets and Facilities Support Services as a Business Resource*. Lambert Academic Press. 2012. (ISBN: 978-3-8465-2785-6).

Wood, B. *Building Care*. Blackwell Science. 2003.

Articles on space planning and workplace management

Brennan, R.J. 'A Paradigm in Transition: The Next Generation of Real Estate and Workplace'. *The Leader*. March/April 2010, pp. 12–18.

Gill, T. J. 'Workplace Continuity: How Risk and Technology Will Affect Facilities Strategy'. *Journal of Facilities Management*. 2006. Vol. 4, No. 2, pp. 110–25.

Harris, R. 'Space – But Not as We Know It'. *Facilities Manager*. 2008. www.i-fm.net/?content/?Article/04/04/081209_04 (accessed 28 Aug 2009).

Olesand, N. 'The Evolving Workspace'. *Premises & Facilities Management*. 2008, pp. 14–16.

Osgood, R. T. *Key Issues in Enterprise Alignment: Survey Results and Practical Experience with the Fortune 1000*. Flad Architects and CoreNet Global. 2009.

Then, D. S.-S. 'The Changing Workplace – Analysing Space Needs'. *Managing Business Support Services*. Reuvid, J. and Hinks, J. Consultant Eds. Kogan Page. London. July 2001, pp. 87–98. (ISBN: 0-7494-3061-3).

——'Post Occupancy Evaluation'. *Best Value in Construction*. Kelly, J., Morledge, R. and Wilkinson, S. Consultant Eds. Blackwell Science. April 2002, pp. 261–79. (ISBN: 0-6320-5611-8).

——'Strategic Management'. *Workplace Strategies and Facility Management – Building in Value*. Best, R., Langston, C. and de Valence, G. Eds. Butterworth-Heinemann. January 2003, pp. 69–80. (ISBN: 0-7506-5150-4).

——'The Business of Space'. *Facilities Change Management*. Finch, E. Ed. Wiley-Blackwell. 2012, pp. 57–75. (ISBN: 978-1-4051-5346-1).

Thomas, D. 'Too Much Space; Too Much Waste'. www.i-fm.net/features/too-much-space-too-much-waste (accessed 28 Aug 2009).

Veale, P. R. 'Managing Corporate Real Estate Assets: Current Executive Attitudes and Prospects for an Emergent Management Discipline'. *Journal of Real Estate Research*. 1989. Vol. 4, No. 3, pp. 1–22.

Zeckhauser, S. and Silverman, R. 'Rediscover Your Company's Real Estate'. *Harvard Business Review*. Jan–Feb 1983, pp. 111–17.

Articles on facilities asset maintenance and renewal

Biedenweg, R. 'Why You Need Life Cycle Planning: The New Breed of Capital Renewal Planning Will Revive Your Campus Even in a Down Economy – On the Money'. *University Business*. FindArticles.com. 20 May 2011. http://findarticles.com/p/articles/mi_m0LSH/is_3_6/ai_98994748 (accessed 24 October 2012).

Booty, F. *Facilities Management Handbook*. 3rd Edition. Butterworth-Heinemann. 2007.

Hardy, V. and Roberts, P. 'International Emergency Planning for Facilities Management'. *Journal of Facilities Management*, 2003, Vol. 2 No. 1, pp. 7–23.

Hardy, V., Roper, K. O. and Kennedy, S. 'Emergency Preparedness and Disaster Recovery in the US post 9/11'. *Journal of Facilities Management*. 2009. Vol. 7 No. 3, pp. 212–23.

Renaud, R. and Phillips, S. 'Developing an Integrated Response Programme for Facilities: the Experience of Public Works and Government Services Canada'. *Journal of Facilities Management*. 2003. Vol. 1, No. 4, pp. 347–64.

Summerell, R. 'Real Property Asset Management Planning'. www.businessfacilities.com/bf_05_07_news2.asp (accessed 16 February 2006).

Thomas-Mobley, L. and Khuncumchoo, N. 'A Facility Manager's Approach to Standardized Construction Contracts'. *Journal of Facilities Management*. 2006. Vol. 4, No. 4, pp. 234–44.

8 Facilities management: a business function

The previous four chapters (Chapters 4–7) have provided a broad outline of the fundamentals of facilities asset management as one facet of facilities management; the other two being the management of space and of services (see Chapter 1). Putting this into practice requires more than knowledge of the fundamentals. It requires strategic and process integration with the management of space and services as well as the necessary support structures that business functions need. This chapter deals with the support structures that are relevant to facilities management.

As a business function, facilities management requires what can be loosely termed as a supporting capability. This includes the combination of policies, strategies, plans, information, systems and other operational tools that enable facilities management to be implemented effectively. Two key aspects of this supporting capability need to be acknowledged. First, the exact nature and format of this supporting capability framework will obviously vary between organizations, tailored to suit the particular organization's strategic direction and its facilities management needs. Second, facilities management is dynamic in nature and subject to change and hence this supporting capability framework must also be adaptable to change since it shapes the practice of facilities management.

8.1 Facilities management and business support

We see facilities asset management as one of the four resource pillars required for the fulfilment of business objectives from strategy to delivery. Hitherto, the contributory role of facilities management has been viewed as mainly operational and its budgetary allocation seen as a cost to business. We hope that we have provided the justifications for the essential, if not critical, business support roles that facilities management plays within an enterprise context in terms of the management of *assets, workspace* and *services* (see Figures 1.3 and 1.4). The context of how the business of managing assets (which is the focus of this book) can add value to prudent business management can best be illustrated in Figure 8.1.

It is worth emphasizing that facilities assets are business resources providing the working environment for the most expensive business resource – your employees.

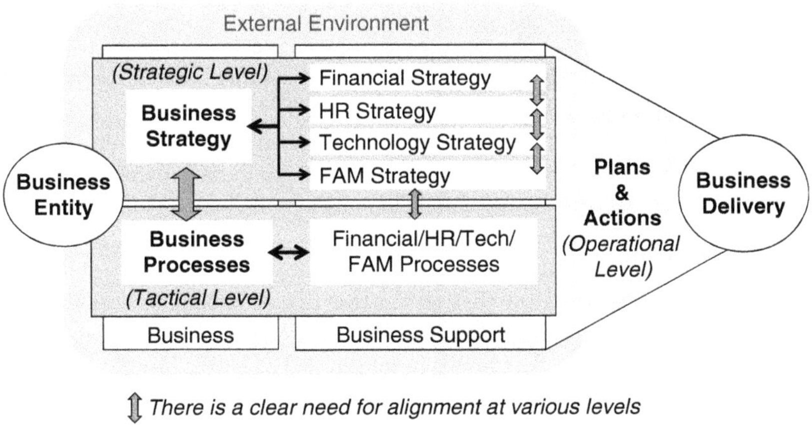

Figure 8.1 Business and business support

How the corporate facilities assets are procured, configured and managed is clearly central to achieving the strategic business objectives. The fact that facilities occupancy costs account for the second or third largest corporate expense for most business enterprises further justifies the potential significant added value that a well-resourced and well-managed facilities management function can bring to the enterprise, given that facilities solutions often cut across functional divisions.

The dynamic nature of business markets within a globally connected environment is particularly challenging for most business enterprises. Since facilities assets are expensive, durable products, relatively inflexible to change and bounded by complex legal rights, their utilization and impact on employees' performance are now key concerns of corporate managers. The management of change in the built environment is a topic of concern not only to governments and the general public; it is increasingly bearing down on issues of global sustainability, corporate social responsibilities and corporate branding. Some of these change drivers include:

- globalization and workforce demographic changes that impact on the workplace expectations, design and use (Agpar 1998, Grantham 2000, Cairncross 2001, Gartner and MIT 2001, Malone 2004, Brennan 2010, Chartered Management Institute 2008, Ernst & Young 2010);
- global concerns about environmental sustainability that are driving the designs of new buildings, the way they are managed in use and expectations of long term performance (Shah 2007, Ratcliffe and Saurin 2008, Hodges 2005, KPMG International 2012);
- the pursuit of effectiveness, in addition to efficiencies through innovations and productivity improvements and their impact on the provision and utilization of workspaces (Lichtman 2006, Cardellino and Finch 2006, Haynes 2007, 2008);

- buildings that are becoming increasingly sophisticated in technology as advances in technology enable them to be designed and built with new materials, systems and operating capabilities (CIC – Pennsylvania State University 2012, Wilkinson and Reed 2009, May and Williams 2012);
- advances in technology that provide facilities management with more effective and efficient systems, tools and information to make informed decisions and provide a better service (Apgar 2009, Valentine and Zyskowski 2009, Ashcraft 2009, Waheed and Fernie 2009, BCIS 2011);
- innovations in the nature of business relationships and communications (e.g. greater use of social media) that affect the way facilities management is perceived, procured, delivers its service value and sustains itself as a vibrant industry into the future (Barnes 2010, Jensen 2012, 2008, Mudrak *et al.* 2005, Kadefors and Brochner 2004, Grimshaw 2003, Pertz 1995, Jones 1995, Barnhoorn 1995).

Given the changing dynamics confronting the role of facilities management in providing the appropriate enabling facilities and supporting services for business enterprises to achieve their goals, it is critical that the practical implementation of facilities management within an enterprise setting are adequate and effective. In this respect, the requirements of effective facilities management are no different from those for the effective management of other business resources like finance, human resources and technology. Key elements of the supporting capability for facilities management discussed in the following sections include:

- business plans
- policies
- organization and resources
- systems
- information
- budgets
- performance measures.

8.2 Key elements of facilities management business support

8.2.1 Business plans

Business plans are comprised of the strategic, operational and tactical plans for the enterprise to advance towards its business objectives and provide the necessary guidance for departments which in turn develop their own departmental or business unit plans. A facilities management department or function also needs a business plan to drive and guide its operations. The only basis for a facilities management business plan is the business it is supporting. The strategic and operational plans for the facilities management department must align with the corporate objectives of the overall business direction, whether strategic objectives involve growth, consolidation or shrinkage.

8.2.2 Policies

Facilities management policies and the standards and guidelines based on them provide the overarching operational direction for the facilities management function and define its role in the organization. For optimum benefit these policies need to be integrated and congruent with business, financial, human resource and technology policies. Standards and guidelines give substance to policies and provide the foundation for their implementation. They provide the detailed guidance for acceptable and unacceptable practices and process outcomes. More importantly they provide the foundations for continuous improvement. Without policies, standards and procedures that show how the facilities management department conducts its business, it would be difficult to assess performance and raise the bar for improvement.

8.2.3 Organization and resources

A robust, effective and efficient facilities management department is fundamental to managing facilities and optimizing their contribution to business outcomes. Regardless of whether the facilities management department is wholly outsourced or in-house supplemented with outsourced resources, the facilities management department needs to be well organized, staffed with competent and motivated professionals and appropriately positioned within the overall business enterprise. A well-defined charter that establishes the role of the facilities management department and its relationships with other resource management heads (i.e. finance, human resources, technology) of the enterprise is important in ensuring that the department is able to respond to changing business strategies and direct its efforts to continuously align facilities provisions and services requirements to changing business needs.

A facilities management department needs adequate resources in terms of the necessary 'hardware' and 'software' in order to perform its roles effectively and efficiently. The 'hardware' is the accommodation, equipment, tools, vehicles, communications, information systems and other supporting facilities. The 'software' includes the appropriate knowledge and competences to evaluate, plan and prioritize in order to derive optimum facilities solutions to meet business demands. Building technology and information technology have advanced significantly in recent years. Building regulations and codes have continued to evolve, becoming increasingly complex and stringent. Users' requirements and expectations have become more demanding, while budget allocations to facilities management are continually being challenged. Without the proper 'hardware' and 'software', a facilities management department is handicapped and cannot function effectively in providing appropriate facilities and services and adding value to business operations and objectives.

8.2.4 Systems

A facilities management system refers to the policies, processes, software and hardware used to capture, store and manipulate facilities-related information, produce management reports, streamline facilities management processes and enhance strategic, tactical and operational decision-making. It is the organizational spine that processes, integrates disparate data, evaluates options, and provides the required information for facilities decisions and solutions. The capabilities and governance of the facilities management system provide the basis for continuous improvement in process efficiency and optimization.

Implementation requires careful planning, organization of processes and clear definition of what is required of the system at all levels. Integration with other systems allows non-facilities-related business information to be accessible to facilities managers and vice versa. At the heart of any facilities management system is the information contained within it. The integrity and reliability of this information are critical to achieving the best value from the investment in a facilities management system.

8.2.5 Information

Access to accurate and reliable information is at the heart of facilities management initiatives. Facilities-related demands do not adhere to departmental boundaries and can emanate from different stakeholders within the enterprise. However, as custodian of the corporate working environment, a facilities management department needs reliable and timely information to be effective and as a sound basis for continuous improvement and benchmarking. The quality and reliability of information are clearly influenced by the way in which the facilities management system is structured, designed and implemented. This is an area that has seen tremendous progress in the application of information technology applications to the whole facilities management process. Nevertheless, it is the information requirements that must drive system design and implementation and not the other way round. This information is not only about the physical aspects of the facilities and their operation and management but must also include access to headcount data from human resources, financial information such as acquisition and leasing costs, asset capital value and depreciation. The organization needs to identify and define the types of management reports and information it requires for senior executives and others in the organization to understand facilities management operations in order to provide performance feedback and policies for implementation. Data collection, analysis and reports to provide timely, relevant and reliable information require significant financial investment and focused management commitment. Resource requirements can be substantial not only in respect of initial establishment but also in terms of ongoing maintenance of data integrity and reliability. This effort should be viewed in the context of the value of the information to effective facilities management.

8.2.6 Budgets

Financial capability in the form of appropriate budgets is clearly a significant factor for all facilities management departments in order to adequately fund facility operations, management, maintenance and capital investment in renewals and new facilities, to meet business operational demands. In reality, facilities management exists in an environment of competing priorities and rarely will there be sufficient funds to do everything that is needed. Business conditions and management decisions will obviously impact on funding capabilities. Peaks and troughs in funding levels are a reality and this dynamic situation is not uncommon in many organizations. Short-term funding shortfalls may not have significant and lasting effects, but a prolonged and chronic constriction in facilities funding often generates liabilities that may require more resources to rectify in the future than would be the case if a more timely response had occurred. For most organizations, the primary purpose of provision for facilities management budgets is to ensure the continued operational functionality of the operational facilities and to preserve the financial worth of the physical assets (particularly if they are assets owned by the organization). It is often not widely known that many non-property or non-construction companies have 20 to 40 per cent of their assets' worth tied up in building assets.

The challenge facing facilities managers and senior business executives is twofold:

- the capability of facilities managers to prepare and defend budget bids that are credible and based on sound arguments supported by quantitative and qualitative information; and
- the ability and willingness of senior executives to develop a sound appreciation of the consequences of budget decisions not only on facilities outcomes but also in terms of their short- and long-term implications on the business.

Senior business executives, with the help of their facilities management executives, can establish (and help to protect) budgets that, notwithstanding dynamics of priorities and business fluctuations, are adequate for operating and maintaining facilities, preserving their useful lives and implementing a proactive, well planned capital renewal and investment programme in line with business requirements.

8.2.7 Performance measures

While the above factors form the basis of a sound business framework for the facilities management function, they are by no means all that is necessary for an effective and efficient facilities management function. Senior management wishing to gain a better understanding of the performance of their facilities management departments in contributing to the enterprise's success would want to know whether (for a given level of investment in facilities management resources) corporate management and customer expectations are being met. Performance measures are important to enable the organization to evaluate facilities management performance and outcomes in the context of its overall business strategy and objectives. An integral part of performance measurement is the use of benchmarks and other comparative tools to drive improvement. Performance measures can also

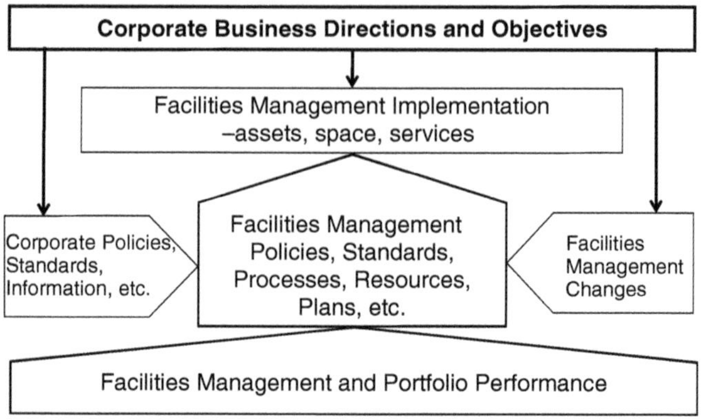

Figure 8.2 Role of facilities management supporting capabilities

be used as part of an audit process. An objective, structured process of a *facilities management audit* is a useful tool in evaluating the effectiveness of facilities solutions and the quality of services, providing clarity of processes and highlighting successful performance and areas for improvement (Whitaker 1995).

Figure 8.2 summarizes the role of supporting capabilities in facilities management implementation. Within an organizational context, the above elements of supporting capabilities will be reflected and influenced by the structure (positioning), processes and competencies of the facilities management function within the enterprise (Then 2000). These aspects will form the basis for the next chapter.

References

Apgar, M., IV. 'What Every Leader Should Know about Real Estate'. *Harvard Business Review*. Nov 2009, pp. 100–07.

Ashcraft, H. 'Building Information Modeling: A Framework for Collaboration'. June 2009. www.scl.org.uk. (Accessed 13 July 2010).

Barnes, R. 'Concept of the Chief Facility Executive'. *International Journal of Facility Management*. April 2010. Vol. 1, No. 1, pp. 1–14.

Barnhoorn, J. G. M. 'Partnerships: A New Management Tool for the Next Decade'. *Facilities*. 1995. Vol. 13, No. 11, pp. 13–18.

BCIS. 'RICS 2011 Building Information Modelling Survey Report'. RICS, July 2011.

Brennan, R. J. 'A Paradigm in Transition: The Next Generation of Real Estate and Workplace'. *The Leader*. Mar/Apr 2010, pp. 12–18.

Cairncross, F. *The Death of Distance*. Boston, Harvard Business School Press. 2001.

Cardellino, P. and Finch, E. 'Evidence of Systematic Approaches to Innovation in Facilities Management'. *Journal of Facilities Management*. 2006. Vol. 4, No. 3, pp. 150–66.

Chartered Management Institute (CMI), UK. *Environmental Scanning – Trends Affecting the World of Work in 2018: Executive Summary*. 2008, pp.12–15.

Ernst & Young. *Business Redefined – A Look at the Global Trends that are Changing the World of Business.* Ernst & Young Global Limited. 2010.

Gartner and MIT. *The Agile Workplace: Supporting People and Their Work.* 2001, pp. 20.

Grantham, C. *The Future of Work.* New York. McGraw-Hill. 2000.

Grimshaw, R. 'FM: The Professional Interface'. *Facilities.* 2003. Vol. 21, No. 3/4, pp. 50–57.

Haynes, B. P. *Office Productivity: a shift from cost reduction to human contribution.* Facilities, 2007. Vol. 25, No. 11/12, pp. 452–62.

—— 'Impact of Workplace Connectivity on Office Productivity'. *Journal of Corporate Real Estate.* 2008. Vol. 10, No. 4, pp. 286–302.

Hodges, C. P. 'A Facility Manager's Approach to Sustainability'. *Journal of Facilities Management.* 2005. Vol. 3, No. 4, pp. 312–24.

Jensen, P. A. 'The Origin and Constitution of Facilities Management as an Integrated Corporate Function'. *Facilities.* 2008. Vol. 26, No. 13/14, pp. 490–500.

—— 'The making of the FM Value Map.' In Jensen, Van Der Voordt, Coenen (eds), The Added Value of Facilities Management: Concepts, Findings and Perspectives. 2012. Polyteknisk Forlag. Denmark. pp. 44–55.

Jones, O. 'Corporate Partnering in Facilities Management'. *Facilities.* 1995. Vol. 13, No. 13, pp. 21–26.

Kadefors, A. and Brochner, J. 'Building Users, Owners and Service Providers: New Relations and Their Effects'. *Facilities.* 2004. Vol. 22, No. 11/12, pp. 278–83.

KPMG International. *Expect the Unexpected: Building Business Value in a Changing World.* 2012. www.kpmg.com/sustainability.

Lichtman, H. *Telepresence, Effective Visual Collaboration and the Future of Global Business at the Speed of Light.* Human Productivity Lab. 2006.

Malone, T. W. *The Future of Work.* Boston, Harvard Business School Press. 2004.

May, M. and Williams, G. The Facility Manager's Guide to Information Technology: An International Collaboration. IFMA. Houston, 2012.

Mudrak, T., van Wagenberg, A. and Wubben, E. 'Innovation Process and Innovativeness of Facility Management Organizations'. *Facilities.* 2005. Vol. 23, No. 3/4, pp. 103–18.

Pennsylvania State University. *Building Information Modeling Planning Guide for Facility Owners.* Version 1.01, CIC – The computer integrated construction research program. Department of Architecture Engineering. May 2012.

Pertz, S. 'Redefining Strategic Facilities Planning'. *Facilities.* 1995. Vol. 13, No. 1, pp. 16–24.

Ratcliffe, J. and Saurin, R. *Towards Tomorrow's Sustainable Workplace – Imagineering a Sustainable Workplace Future.* Johnson Controls Facilities Innovation. 2008.

Shah, S. *Sustainable Practice for the Facilities Manager.* Blackwell Publishing. 2007. (ISBN: 1-4051-3557-3).

Then, D. S.-S. 'The Role of Real Estate Assets in Supporting the Fulfillment of Corporate Business Plans – Key Organisational Variables for an Integrated Resource Management Framework'. *Facilities.* 2000, Vol. 18, No. 7/8, pp. 273–80.

Valentine, E. and Zyskowski, P. 'Building Information Models (BIM): How It Has Changed FM'. *Facilities Management Journal.* May 2009. Vol. 19, No. 3, pp. 28, 30–32.

Waheed, Z. and Fernie, S. 'Knowledge based facilities management'. *Facilities*. 2009. Vol. 27, No. 7/8, pp. 258–66.

Whitaker, M. 'Conducting a facility management audit', *Facilities*. 1995. Vol. 13, No. 6, pp. 6–12.

Wilkinson, S. J. and Reed, R. 'Green roof retrofit potential in the central business district'. *Property Management*. 2009. Vol. 2, No. 5, pp. 284–301.

9 Facilities management organization

Previous chapters have described the fundamental aspects of facilities asset management and the supporting capabilities that are necessary to translate facilities management into practice. In this chapter, we elaborate on facilities management organization, which refers to the manner in which the management and administrative reporting structure, resources and other supporting capabilities are assembled and structured to become an effective and efficient business function that is appropriately positioned within the overall enterprise organizational structure. The word 'enterprise' is deliberately used in this section to minimize confusion between the organization of facilities management and the parent organization (enterprise) that it serves.

Being effective means being in alignment with business vision, mission, strategy and objectives and fulfilling the designated mission, roles and responsibilities. Being efficient means fulfilling the designated mission, roles and responsibilities with optimum economy in the use and consumption of business resources. The organizational framework integrates management and administrative structures, supporting capabilities and service models into a cohesive whole and transforms the facilities asset management concept into a fully operational business function within the enterprise.

9.1 Scoping facilities management

The practice of facilities management within business enterprises has been around for almost three decades and it is a growing professional discipline offered by universities in developed and developing countries around the world. As a young discipline, its body of knowledge has evolved to reveal a complex field in which facilities solutions (hard and soft) to business challenges inevitably transcend traditional disciplines like management, finance, economics and human relationships. The potentially broad scope of activities that can be included under the 'umbrella' of facilities management (see Figure 9.1) and the varying degree of maturity of the industry in different countries have resulted in different 'interpretations' of the scoping of facilities management activities. However, even as facilities management professional institutions and academics in Europe, North America and Asia are grappling to come up with a common coherent definition, a consensus has

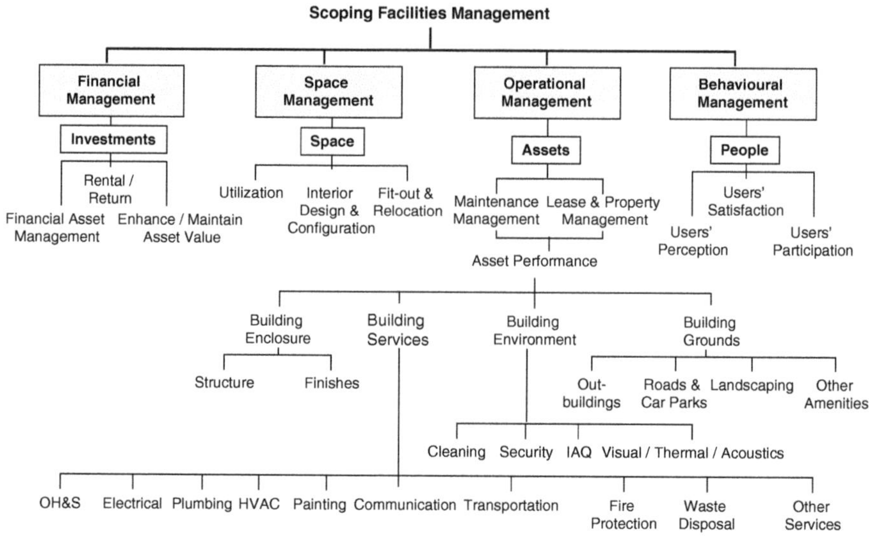

Figure 9.1 Scoping facilities management (Modified from Quah 1998)

emerged that views facilities management as integrative, cross-disciplined and both strategic and operational in practice (Then 2000, 2003; Nutt 2000, Varcoe 2000, Jensen 2008, 2011).

The addition of the words *investments*, *space*, *assets* and *people* to the original diagram by Quah (1998) points to the emphasis on the business resources dimension, rather than the transactions arising from the operational dimensions of facilities management. There is a clear trend in recent years that supports the proposition that the push by businesses to demonstrate value for money necessitates a management approach that embraces all of a business's resources and emphasizes effectiveness on top of general economic efficiency. In this respect, the management of facility assets as a business resource is no exception.

Our scoping of facilities management (FM) under *assets*, *workspace* and *support services* (Figure 1.3) necessitates fundamental shifts in the current conceptualization of FM as essentially an operational response to business dynamics, in terms of moving:

- FM beyond transactions – *about people, investments, assets and space.*
- FM beyond buildings – *about effective deployment and utilization of functional space.*
- FM beyond cost – *about alignment of supply to demand, value contribution to core business, affordable and appropriate services, and whole-life asset management.*

(Then 2012)

The above shifts are also largely in line with developing trends of the FM industry over the last two decades. A scan of learned papers in facilities management journals over the last two decades will reveal that the general development of FM thinking has evolved over four distinct phases (Jensen, Voordt, Lindholm, Riratanaphong and Pfenninger 2012):

- FM merely considered as an overhead to be managed for minimum cost rather than optimum value;
- FM as an integrated continuous process in relation to the organization's core business;
- FM as resource management concentrating on managing supply chain issues associated with the FM functions;
- FM as strategic management to ensure alignment between organizational structure, work processes and the enabling physical environment according to the organization's strategic intent.

A key theme across the four phases of change is a shift from preoccupation with tactical concerns to addressing strategic concerns. Key factors that drive the changing focus in facilities management include (Then 2012):

- dynamics of business environment – *need to cater for the speed of change*;
- demand for flexibility – *need to incorporate built-in exit strategies in facilities provision*;
- demand for global connectivity – *investments in technology and communication infrastructure*;
- changes in work and workplaces – *need to cater for employees' diversity and mobility*.

It is clear from the above that the role definition of the facilities management function within an enterprise is predicated on change which necessitates abilities to respond to the enterprise's organizational change, to predict facilities changes and potential changes to service delivery models. Facilities managers should regard this as an opportunity to promote themselves as a catalyst of change within their enterprise, as any changes in the corporate strategic business direction are likely to impact on some facilities dimensions of facilities management.

From an organizational perspective, two main categories of roles are discernible from the range of activities associated with facilities management:

- facilities provision – activities related to the role of providing the appropriate facilities assets portfolio (i.e. buildings) required by the enterprise to carry out its core business activities; and
- facilities support services management – activities related to the role of the ongoing management and servicing of operational facility assets (i.e. the workplace environment within buildings-in-use).

The distinction between facilities provision and facilities support services management is important in that they are inextricably linked and share the common

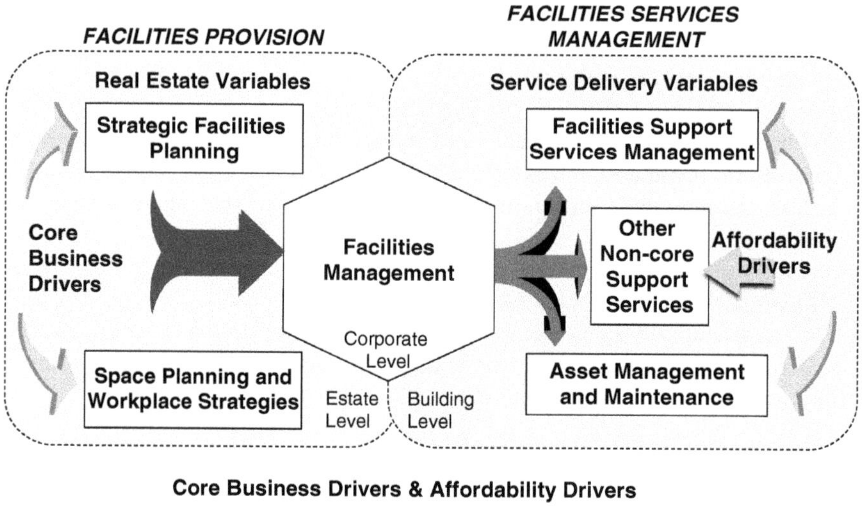

Figure 9.2 Critical interfaces within facilities management (Then 1998)

objective of providing the required enabling workspace environment to fulfil corporate objectives. More importantly, for many enterprises, two or more functional departments or divisions may separately manage the roles of facilities provision and facilities services management. Successful facilities resource management should consider these two facets as one overall issue. In this respect, it is critical to understand the interfaces between the two interrelated processes.

Figure 9.2 illustrates the crucial relationship between strategic decisions (driven by core business drivers) relating to facilities provision issues and tactical–operational decisions (driven by affordability drivers) relating to facilities support services procurement and performance management of the facilities assets portfolio. The core business drivers relate to the facility asset issues of demand evaluation, change management and the appropriate facilities variables needed to support the achievement of corporate objectives. The affordability drivers relate to supply evaluation, procurement strategy and performance monitoring of servicing the workspace within the facilities asset portfolio. It is these critical interfaces that must be catered for in the structural setting of the facilities management roles within the enterprise.

Figure 9.2 also depicts the traditional divide between real estate variables (seen as the domain of corporate real estate management – CRE) and support services delivery (seen as the domain of facilities management – FM). In our opinion, CRE and FM are two sides of the same coin and as such serve the same objective of enhancing corporate performance. This view is largely in line with the finding of a recent survey by Jones Lang Lasalle (advance report 2011) which concluded that, increasingly, best-in-class performers are moving from the traditional silos of transaction management, facilities management, portfolio administration and project management to a more integrated delivery of customer relationship management (CRM), workplace and process management.

In terms of implications for FM competencies, four core components become apparent covering both strategic and operational aspects of facilities management activities:

- Strategic facilities planning – *responding to business dynamics.*
- Space planning and workplace strategies – *workplace design and management.*
- Asset management and maintenance – *asset value and performance.*
- Facilities support services procurement and management – *services delivery.*

The need for an integrated management approach is critical in that the prevailing business strategic direction will drive strategic facilities planning, the outcome being an agreed workplace strategy which in turn feeds into the operational delivery of facilities support services, and asset management and maintenance.

9.2 Requirements of an integrated management framework

The proactive management of the facilities resource demands clear strategic direction from senior management and clear measurable delivery from operational facilities management (see Figure 1.9). In recent years, it has become apparent that there are clear shifts in focus as the practice of facilities management gradually matures. It is already discernible from published literature and review of practices that the initial preoccupation with tasks and functions has given way to an emphasis on processes and their management. More recently, the shift has been towards resource integration, with the emphasis on provision of enabling working environment where the issues of people, processes, technology and property are elements of the same problem seeking a common solution (Figure 9.3).

The changing focus in facilities management as an integrated resource management framework spells dramatic shifts in competencies for both the demand side (purchasers of facilities and services) and the supply side (suppliers and support services contractors/providers). The trend towards organizational 'downsizing' and 'outsourcing' collectively imposed an onus on many enterprises to

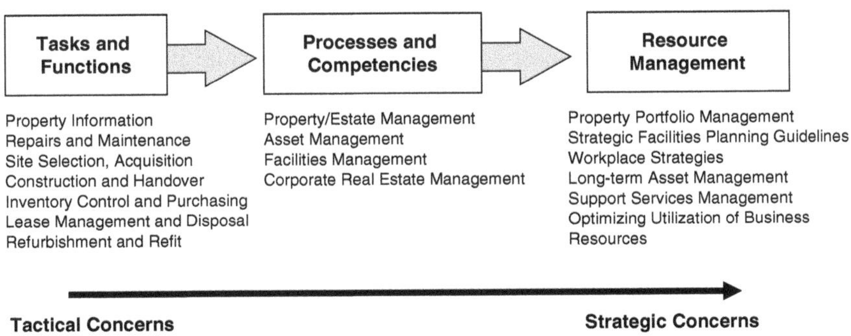

Figure 9.3 Changing focus in facilities management (Then 1998)

Table 9.1 Trends and impact on facilities provision and facilities support services management

Trends in facilities provision	*Impact on facilities provision*
• Growing awareness of need to manage physical resource (mix and fit evaluation) • Concerns with rising occupancy costs • Long-term asset ownership and liabilities • Need to align real estate resource to strategic direction • Environmental concerns (site and internal configuration) and employee productivity	Flexibility in: • Real estate portfolio profile (level of ownership and/or lease liabilities) • Optimizing location advantages (rationalization and disposal) • Space utilization (innovations) • Layout configurations (functional/ process)
Trends in facilities services management	*Impact on facilities support services management*
• Becoming generalist • In search of economy • Trend towards outsourcing • Partnering and alliances • Service orientation	Flexibility in: • Accommodating changes in work style due to technology – ITC, electronic mail, Web, etc. • Alternative procurement strategies • Contracts and service management

Source: Then 1998

seriously review their internal competencies necessary for managing in a new era of choices and flexibility. In this respect, the scoping of activities covered within facilities management and how they are procured, managed and measured have become matters of corporate concern.

Table 9.1 provides a summary of the main drivers pushing change by listing the emerging trends and their implications for facilities provision and facilities services management.

The desired outcome in physical form is an appropriate facilities asset portfolio structure that is aligned with the enterprise's business operational requirements. Buildings and land, as physical assets, are relatively static products. Effective matching of demand for, and supply of, functional accommodation and associated facilities support services to meet operational requirements in a dynamic business environment demands the management of the facility assets as a dynamic integrated process. The realization of an integrated approach necessitates a formal planning framework that must cater for the cultural, procedural and existing knowledge base of the enterprise concerned.

Designing or choosing the right organizational model has to take account of a range of factors, as each enterprise operates in very different ways. A successful team structure can be delivered through a number of platforms with the varying size and structure of businesses leading to a range of hybrids. There have been several studies looking into effective organizational structure for corporate real estate (CRE) and facilities management (FM).

Miciunas (2002) suggests that the CRE/FM organizations must consider the relationship of four strategic design issues: relevance to the enterprise, process and measurement, knowledge and information, and credibility and accountability. The Jones Lang LaSalle (advance report 2011) suggests that a clear evolution of organization changes is already under way with an emphasis on being leaner, more strategic and focused on the internal stakeholder, working alongside the wider business as influencers and leaders of change. However, the report also cautions that such an evolution is likely to require a fundamental reassessment of team structure, roles and skills.

In his research, Then (1998) revealed that performance of CRE/FM departments can be evaluated in terms of three key organizational variables:

- structure – organizational set-up for facilities provision and facilities services management; key factors include positioning, influence and advice;
- processes – the systems and procedures for the management of the delivery of facilities assets and their associated facilities support services; key factors include efficiency, effectiveness and added value;
- competencies – the necessary skills required for an efficient and effective delivery system (both in-house and bought-in expertise); key factors include appropriate skills, market knowledge and providing solutions.

The above three organizational variables of *structure*, *processes* and *competencies* must work in concert, since any mismatch between these variables is likely to lead to a situation where the solutions generated are sub-optimal in the utilization of the facilities assets. The three organizational variables provide a basis for measuring

Table 9.2 Strategic shifts required in structure, processes and competencies

Structure *From:*	*To:*
• Distance from senior management • Low/non influential • Functionally divided • Predominantly in-house service provision	• Close to strategic business planning • Contribute to strategic options • Projects-driven • Competitive procurement embracing outsourcing
Processes *From:*	*To:*
• Reactive, supply-led process • Subservient role, transaction-led • Functionally split, property focused • Patchy knowledge of assets base	• Proactive, demand-led process • Supporting strategic intent, business process-led • Integrated resource management • Interactive knowledge of assets base
Competencies *From:*	*To:*
• Technical skills • Solving problems • Task-based approach • Internally focused	• Business skills • Providing strategic options • Process-based approach • External comparison (benchmarking)

how effectively an organization responds to the demands for changes to its operational facilities asset base as it moves from one steady state to the next over time. Table 9.2 summarizes the necessary shifts in organization structure, management processes and staff competencies required in moving from a traditional reactive management style to a dynamic, proactive, integrated resource management approach.

At another level, globalization is challenging multinational corporations to integrate their business models across mature and emerging markets. The pace of globalization has increased exponentially, creating a need for greater consistency in operations, which in turn demands better quality communications, strategic planning and efficiencies in how basic business resources are managed. Outsourcing has joined globalization and technology to become a primary driver of the need for change in terms of organizational set-up for large multinational corporations with distributed facilities assets portfolios (Jones Lang LaSalle 2009).

The findings of two recent international surveys (Evans 2012, CoreNet Global 2008) point towards a trend of a smaller core team of a strategically focused and centralized CRE/FM department with increasing outsourcing of functions like construction management, workplace design, project management, transaction management and services delivery. Strategic planning of facilities provision, workplace strategies and procurement options evaluation have become key internal functions and have partly accounted for CRE/FM's ascendancy in the corporate hierarchy. Key drivers cited for this shift include:

- the need for a strategic approach – having a global oversight offers the ability for executives to identify strategic opportunities across different business units or geographies;
- driving consistency – notwithstanding local differences, improving consistency in communication systems, processes and reporting will provide a greater level of predictable outcomes, decreasing risk and increasing control;
- efficiencies and speed to market – need for fast decision-making and agility to act quickly as a key driver towards a more integrated approach that provides streamlined structure and process.

However, the benefits to be derived are subject to executive-level sponsorship, CRE/FM taking on an 'intelligent client' role, robust governance structures and adoption of strategic service partnerships supported by effective technology and communications infrastructure.

9.3 Enterprise and external factors

The brief review of requirements of a proactive management framework for facilities management in the previous section points to the need to consider not only internal factors to the enterprise but also external factors that impact on the supply chain for facilities management activities.

9.3.1 Nature of enterprise and facilities portfolio attributes

It would be very rare for facilities management functions to be able to design their own organization framework without due consideration of the objectives and targets set by the corporate leadership/board. After all, the facilities management function is part of the enterprise. How the enterprise is organized (divisions, regions, international and national, etc.) will shape the way in which facilities management is organized and deployed. The existing mandate and remit of the facilities management function should be clear in terms of control, nature and frequency of interaction with corporate management and business units and the number of stakeholders and their input to facilities-related decisions. With outsourcing, the mandate and remit of any service provider also becomes critical as it directly impacts on the resourcing and focus of the in-house team. Other organizational and business issues impacting on facilities management organization include:

- the composition of the business – *number of business units*;
- business environment and industry trends – *degree of volatility*;
- strategic business directions and objectives – *expansion, contraction, etc.*;
- corporate and facility-related policies – *ownership, location, etc.*;
- culture and organizational politics – *role of facilities and workspace.*

The positioning of the facilities management function within the enterprise organization structure relative to other resource heads like finance, human resources and technology is a subject that has attracted much discussion and even research (Nutt 1999, Chotipanich 2004, 2008), given corporate asset worth, high facilities occupancy expenses and the impact of inappropriate facilities on users that directly affects corporate financial performance. Clearly, the actual positioning of the facilities management department within the enterprise organizational hierarchy will have implications for its strategic, operational and tactical roles in terms of communication and influence and, ultimately, resource allocation. The positioning within the organization structure will be reflected in the reporting arrangements: the span of management levels between senior corporate management and the facilities management function. The structural positioning of the facilities management function in relation to other business units and divisions of the enterprise also impacts on its degree of influence and relationships with other parts of the business. The second dimension is the extent of involvement by the facilities management function in enterprise-wide strategic business matters with implications for facilities and vice versa. The extent of strategic involvement may be enhanced through its positioning in the organization structure. However, involvement may also occur in spite of this, due to the third dimension of relationships. Collaborative relationships enable the facilities management function to optimize its strategic role through its position in the management structure or conversely to overcome any difficulties arising from its position in the management structure. In this respect, the nature and extent of dialogue between all stakeholders involved – formal and informal – are critical to facilitate communication both horizontally and vertically within the enterprise.

The nature, extent and complexity of the facilities portfolio also influence the design of the organization framework. The criticality of the facilities to business process influences the perception of and priority placed on facilities management and its organizational framework. Large-scale manufacturing or research facilities with complex building systems will demand a higher level of technical and management competency concentrated in a particular location. A portfolio of schools dispersed over a large geographical areas, on the other hand, would require a more distributed deployment of resources, different management structure and different levels of skills. At a regional or global level, multinational corporations with facilities assets in different countries will face a different set of factors in deciding on the optimum organizational choice. The trend towards outsourcing and shrinking to the core will no doubt feature prominently as a key factor in the structuring of facilities management departments in terms of strategic functions and headcount. Decisions on the extent of outsourcing of facilities management functions or parts thereof would clearly be influenced by the nature and size of the portfolio and how mission-critical the facilities are.

9.3.2 Maturity of facilities management supply market

The operational demands of the business enterprise (nature of business, size, operational requirements, culture, context, etc.) will determine the role and mission of the facilities management function. The general push for economy in the use of business resources and an increasingly sophisticated and maturing facilities service providers' market have meant that outsourcing of in-house functions now represents a compelling economic choice (Jones Lang LaSalle 2009). The service-delivery model adopted by the facilities management function will impact on the size and set-up of its organizational framework in terms of the extent of outsourcing. The demarcation between in-house and outsourced functions and services is influenced by factors such as how mission-critical the services are, corporate policies on outsourcing, the depth and capabilities of the service provider market and recruitment and retention issues. Other factors that influence the choice of service-delivery model include past performance of the facilities management function, industry trends and attitudes towards risk.

For a service provider company that specializes in the provision of facilities management services to other organizations as a commercial service, the purpose and mission of the facilities management function are to generate a profit and therefore it is the company's core business. To achieve this, it applies various strategies to ensure that its business is sustainable, achieves performance targets and generates value to its shareholders. The organizational framework is therefore designed to align with these business objectives.

A facilities management function that operates within an enterprise as a supporting function has a different purpose and mission. Its organizational framework would be designed to match the operational requirements of the business. Different variations of the 'supporting function' model would obviously exist in both private and public sector enterprises. In order for the facilities management function to add value to the enterprise and contribute to the achievement of its strategies and objectives, a clear understanding of the business of the enterprise is necessary. The scale and nature of the business will define the purpose and mission of the facilities management organization so that it is aligned with the business strategy and objectives, and guide the structure of its organizational framework.

9.3.3 Facilities management's mandate and mission

The purpose and mission of the facilities management function are key factors in the design and structure of its organizational framework. They define the roles and responsibilities of facilities management at strategic, operational and tactical levels and how the human resources, systems, policies, information, plans and processes should be organized to provide an efficient and effective service. As the custodian of the business working environment, the facilities management function is increasingly seen as essential and critical in enabling the most expensive business resource – its staff. The physical workplace environment within business enterprises is increasingly acknowledged as a catalyst for change and a reflection of corporate culture. In this respect the role of facilities and how they are provided, serviced and managed have become strategic issues as parts of the successful implementation of business strategies and achievement of business objectives. The purpose and mission of the facilities management function should, amongst other things, reflect this crucial role.

It would be impossible to identify all the possible factors to be considered, given that each business operates in different ways, but in each case there are common factors which need to be addressed. Figure 9.4 is an attempt to summarize some common factors influencing the design of facilities management organization

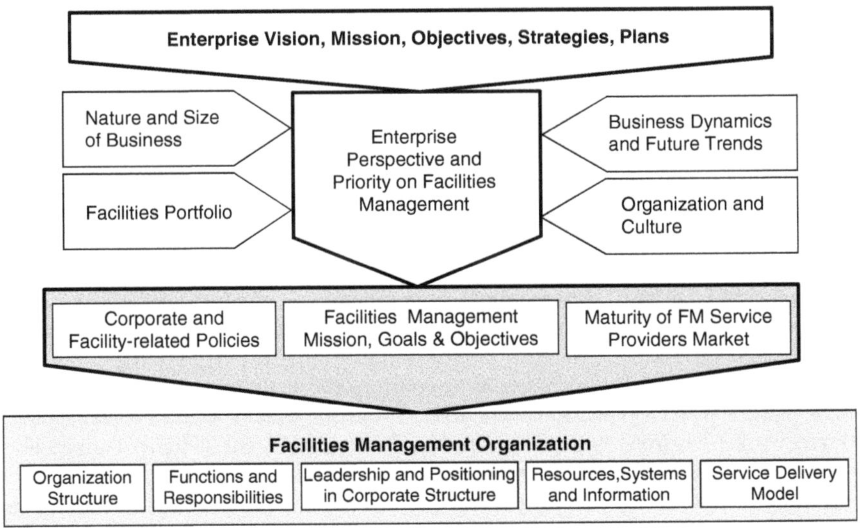

Figure 9.4 Facilities management organization design

within enterprises. There is strong justification that, provided the right people, systems, technology and processes are in place, the successful implementation of facilities management organization within an enterprise is not only possible but inevitable.

References

Chotipanich, S. 'Positioning Facility Management'. *Facilities.* 2004. Vol. 22, No. 13/14, pp. 364–72.

Chotipanich, S. and Nutt, B. 'Positioning and Repositioning FM.' *Facilities.* 2008. Vol. 26, No. 9/10, pp. 374–88.

CoreNet Global Applied Research Center. 'Corporate real estate organizational structure and design trends.' A CoreNet Global Applied Research Bulletin, in partnership with Deloitte Consulting. October 2008.

Evans, B. 'A Recipe for Success: The Globalization of Real Estate Management Structures.' *The Leader.* CoreNet Global. Mar/Apr 2012, pp. 12–18.

Jensen, P. A. 'The Origin and Constitution of Facilities Management as an Integrated Corporate Function.' *Facilities.* 2008. Vol. 26, Nos. 13/14, pp. 490–500.

——'Organisation of Facilities Management in Relation to Core Business.' *Journal of Facilities Management.* 2011. Vol. 9, No. 2, pp. 78–95.

Jensen, P. A., Voordt T., Lindholm, A., Riratanaphong, C. and Pfenninger, M. 'In Search for the Added Value of FM: What We Know and What We Need to Learn. *Facilities.* 2012. Vol. 30, No. 5/6, pp. 199–217.

Jones Lang LaSalle. 'Perspectives in Organizational Design.' Authors: McBliane, R., McCarty, T., Schuth, J. and Gottschalk, D. Feb 2009.

——'Better by Design: Reshaping the CRE Function for Greater Impact.' Authors: Bayne-Jardine, T., Frost, S., Schuth, J. and Carroll, T. Apr 2011 (advance report).

Miciunas, G. 'What Makes for Effective Organizational Structure in Corporate Real Estate? Going Beyond Reporting Structure and Sourcing Decisions to Considering Strategic Design Issues.' *Journal of Corporate Real Estate.* 2002. Vol. 5, No. 1, pp. 19–30.

Nutt, B. 'Linking Facilities Management Practice and Research'. *Facilities.* 1999. Vol. 7, No. 1/2, pp. 11–17.

——'Four Competing Futures for Facility Management'. *Facilities.* 2000. Vol. 18, No. 3/4, pp. 124–32.

Quah, L. K. Ed. *Facilities Management and Maintenance: The Way Ahead into the Millennium.* Proceedings of CIB W70 International Symposium on Management, Maintenance and Modernisation of Building Facilities. Singapore. McGraw-Hill Book Co. 18–20 Nov 1998, p. xv. (ISBN: 978-9971-8997-69).

Then, D. S.-S. 'Corporate Leadership in Real Estate and Facilities Management.' Proceedings of CIB W70 International Symposium on *Facilities Management and Maintenance: The Way Ahead into the Millennium.* Singapore. Quah, L. K. Ed. Nov 1998, pp. A9–A16.

——'The Role of Real Estate Assets in Supporting the Fulfillment of Corporate Business Plans – Key Organisational Variables for an Integrated Resource Management Framework.' *Facilities.* 2000. Vol. 18, No. 7/8, pp. 273–80. (Highly Commended Award 2001 by LITERATI club.)

——'Integrated Resources Management Structure for Facilities Provision and Management.' *Journal of Performance of Constructed Facilities.* Feb 2003. Vol. 17, No. 1, pp. 34–42.

——'Scoping Facilities Management – An Inclusive Practice View.' Paper presented at the IFMA Integrate Conference, Hong Kong. 5–6 June 2012.

Varcoe, B. 'Implications for Facility Management of the Changing Business Climate.' *Facilities.* 2000. Vol. 18, No. 10/11/12, pp. 383–91.

Other useful reading

Alexander, K. Ed. *Facilities Management.* E & FN Spon. 1996.

Cotts, D.G. and Lee, M. *The Facility Management Handbook.* AMACOM.1992.

Friday, S. *Organization Development for Facility Managers.* AMACOM. 2003.

Kaiser, H. *The Facilities Manager's Reference.* R.S. Means Company. 1989.

Kaya, S. and Alexander, K. 'Classifying FM Organisations Using Pattern Recognition.' *Facilities.* 2005. Vol. 23, No. 13/14, pp. 570–84.

——'Classifying Client Side FM Organisations in the United Kingdom: Why My In-house FM Organisation is Dissimilar to Competitors and Others.' *Journal of Facilities Management.* 2006. Vol. 4, No. 2, pp. 86–98.

Kaya, S., Heywood, C. A., Arge, K., Brawn, G. and Alexander, K. 'Raising Facilities Management's Profile in Organisations: Developing a World-class Framework.' *Journal of Facilities Management.* 2005. Vol. 3, No. 1 pp. 65–82.

McGregor, W. and Then, D. S.-S. *Facilities Management and the Business of Space.* Arnold. 1999; reprinted under Butterworth-Heinemann. 2001. (ISBN: 0-340-71964-8).

Pitt, M. 'Total Business and Facilities Management: Implementing a Holistic and Dynamic Self-re-enforcing Framework for Business Certainty, Continuity and Growth.' Editorial. *Property Management.* 2005. Vol. 23, No. 1.

Then, D. S.-S. 'Integration of Facilities Provision and Facilities Support Services Provision – A Management Process Model.' Proceedings of the 2002 Glasgow CIBW070 Global Symposium – Exploring the Global Knowledge Base on Asset Maintenance Management, Workplaces and Facilities Management, Scotland, 18–20 Sep. CIB Publication No. 277. 2002, pp. 38–52.

——'Concepts in Facilities Management.' *Architect's Handbook of Construction Project Management.* Murray, M. and Langford, D. Eds. RIBA Enterprises. 2004, pp. 297–312. (ISBN: 1-85946-123-9).

——*Real Estate Asset Management – Operational Property Assets and Facilities Support Services as a Business Resource.* Lambert Academic Press. 2012. (ISBN: 978-3-8465-2785-6). (Doctoral thesis. 1996. Heriot-Watt University, Edinburgh.)

10 Facilities management alignment to business needs

We hope that we have provided the justifications for the growing international recognition of the important role of building facilities as a business resource that has led to a greater awareness of the need to manage facilities assets more effectively. This recognition has also driven a change in perception of facilities management (FM) towards that of a management function that can contribute to business success. Corporate awareness of the importance of workplaces and their associated occupancy costs in recent years has enhanced the role of facilities asset management in both the private and public sectors around the world (Becker and Kelley 2004; McGregor and Then 2001; Carter 1999; Hastings 2010). Previous chapters have dealt specifically with facilities asset management as one of the facets of facilities management and the central theme of this book. In this chapter we expand our field of vision to the wider scope and application of facilities management in the context of a management function.

Globalization, technological advances and shifting management priorities have altered traditional views of real estate facilities and workspace provision. Improved communication has increased inter-country and workplace mobility and resulted in changes to traditional supply chains. These changes, nationally and globally, have altered both the assessment and procurement of business support infrastructure, including the provision and management of facilities assets. To this end, organizations have pursued various strategies aimed at better matching supply with demand: all efforts aimed towards better control and management of facilities occupancy costs. In the practice of facilities management, there has been a greater awareness and realization of the need for integrated resources planning (i.e. people, finance, property and technology) in order to derive optimum facilities solutions to meet emerging business challenges (Chapter 3). A central theme of any business optimization process is the concept of alignment of business resources to meet changing business needs.

10.1 Need for alignment between business needs and facilities management strategies

The need for alignment between business needs and the organization's facilities infrastructure is at the heart of any strategy in supporting business success. All businesses need real estate facilities and facilities services in order to function – whether that business is a manufacturing company, a financial institution, a retail outlet, a hospital or a school. Matching real estate facilities and facilities services to business needs is an obvious requirement in terms of economic resource consumption and utilization, whether for profit-driven or not-for-profit organizations. The basic premise is the acknowledgement that business performance is dependent on having the appropriate operational facilities and services to support business delivery (Lindholm and Leväinen 2006, Then 2004, Krumm and de Vries 2003).

Strategies to improve utilization of business facilities assets and facilities services performance have been influenced and impacted by:

- increased demand for exploiting technology and knowledge management applications in the design and management of modern facilities and workspaces (Robertson 2000; McGregor and Then 2001; Joroff 2002; Ratcliffe and Saurin 2008; IBM Institute of Business Value 2010);
- widespread outsourcing of FM services, to varying degrees and using a number of procurement models (Barnhoorn 1995; Frost 1997; McBlaine and Moritz 2002);
- increased focus on facilities assets as a business resource and methods for measuring facilities and workspace performance that reflect business goals and corporate branding (Varcoe 2002; McDougall *et al.* 2002; Then 2003; Valence 2004).

Facilities management by nature is a dynamic business activity and must be responsive to changing client needs in order to fulfil its increasingly critical role as custodian of the corporate workplace and workspace environment. The extent to which these responses are reactive or planned (as an FM service anticipating the needs of the client and responding in a timely and cost-effective manner) varies from organization to organization. This variation is influenced by the nature of the business, corporate perception of the role of real estate facilities, the competence of the in-house facilities management team and/or external service provider(s) and other factors.

More recently, research and practice directed at understanding the role and performance of facilities assets in enabling core business strategy is receiving increasing attention (Then and Tan 2002; Nutt 2004; Osgood 2004, 2009; Green and Jack 2004; McDonaugh and Nicola 2009). In a recent longitudinal survey study of Fortune 1,000 companies, Osgood (2009) noted obvious relationships between core business and facilities/real estate elements. However, the exact nature of these relationships is not well understood or thoroughly researched. It is against this background that the concept of alignment (Kaplan and Norton 2006) is being applied to the role of facilities management within an organizational context.

In facilities management, the need for alignment exists at different levels (see Figure 8.1) and is driven by the process of continuous matching of supply to demand, both in terms of real estate facilities and facilities support services and intangibles such as management processes, leadership and competencies.

10.2 Alignment and its application in facilities management

The concept of alignment within enterprises first surfaced with the book *Alignment* by Kaplan and Norton (2006), based on their original works on 'the balanced scorecard' and strategies to improve overall corporate performance (Kaplan and Norton 1992, 1996, 2000). Kaplan and Norton (2006) aptly summarized alignment as a source of economic value in the following quotations:

> "When the enterprise aligns the activities of its disparate business units and its support units, it creates additional sources of value … " (p. 5).
>
> "Corporations must continually search for ways to make the whole more valuable than the sum of its parts. Alignment is critical if enterprises are to achieve synergies throughout their business and support units … " (pp. 26–27).
>
> "Alignment is NOT a one-time event … " (p. 245).
>
> "By its very nature, alignment requires cooperation across organization boundaries, and therefore the process must be managed proactively … " (p. 257).

We first examined the issue of alignment by exploring the potential impact of FM performance on business performance (Then and Tan 2006). The central theme of the research is to better understand the variables underpinning alignment of FM as an integral part of fulfilling business needs. The research started with a focus on an analysis of the dynamics between the role of FM and the role of supporting facilities assets in meeting business needs. The proposition is that the synergies achieved through the alignment of the relevant parameters create an FM function that is optimally synchronized with the needs of the enterprise – it is neither more nor less than what is precisely required to support the business enterprise. Conversely, symptoms of misalignment would include situations within businesses in which there is a mismatch between demand for and supply of appropriate workspace, services delivered within business facilities are inadequate to meet operational requirements, or resources allocated to facilities management are inappropriate for efficient and effective business delivery and operate in conflict with corporate policies and culture. Understanding of the extent of alignment (or misalignment) of facilities provision and facilities services with business needs will assist in:

- improving the positioning of facilities assets and services to be more effective in supporting the achievement of corporate objectives;
- managing the dynamic nature of FM and its strategic and operational responses in order to align with the corporate direction of the business and other external factors;
- focusing FM resources and expertise to have a better understanding of business operational needs and drive more effective utilization of the corporate facilities asset portfolio; and
- aligning FM performance with appropriate performance indicators aimed at measuring the cost and value contribution of facilities assets and facilities services to business outcomes.

10.3 A business–facilities management alignment framework

Conceptual development of the alignment framework was guided by, and based on, a number of contextual propositions which provided the link between the concept of alignment as a key to business success and its application to the realm of facilities management practice. Table 10.1 summarizes the seven propositions which guided the development of the Business–FM alignment framework.

The context of FM alignment to business (Proposition 6) is summarized in Figure 10.1 which clearly affirms that the source of facilities solutions is business objectives and strategies.

Figure 10.2 is a conceptual representation of Proposition 7. Figure 10.2 conveniently delineates the context of the business domain and that of the FM domain. Taking a horizontal delineation, the top segment of the figure comprises the business domain in which *business needs* drive *facility solutions* and the allocation of *FM resources* through the FM strategy, which is shaped by the

Table 10.1 Facilities management alignment propositions

Proposition 1	FM is a vital support function to the enterprise, similar to HR, Finance and IT.
Proposition 2	Support functions do not have a direct role in delivery of core business services but have a role in contributing to competitive advantage of the enterprise.
Proposition 3	Support functions contribute to competitive advantage by providing efficient and effective support services that align with enterprise strategies.
Proposition 4	An effective support function 'understands its customers' strategy and uses its functional expertise to create and deliver solutions that contribute to its customers' success' (Kaplan and Norton 2006, p. 139).
Proposition 5	Alignment of FM with enterprise strategies is critical to the efficient and effective provision of FM services and facilities solutions that add value to the enterprise and enhance its competitive advantage.
Proposition 6 (Refer to Figure 10.1)	When alignment exists, the FM function understands the business strategies of the enterprise, including the component business units (internal customers of the FM function) and develops its facilities strategies, service portfolio, resources and business processes to create and deliver facilities solutions that support its customers and stakeholders.
Proposition 7(Refer to Figure 10.2)	Based on Proposition 6, our concept of FM alignment can be expressed in terms of four dimensions or variables of alignment: Variable 1: Supply and demand alignment Variable 2: FM service alignment Variable 3: FM resource alignment Variable 4: Organizational alignment.

organization's business strategic direction. The bottom segment defines the FM domain, within which the appropriate facility solutions, FM resources and FM services are delivered to support the achievement of the organization's business objectives through its FM strategy.

Figure 10.1 Context of alignment of facilities management to business (Then and Tan 2010)

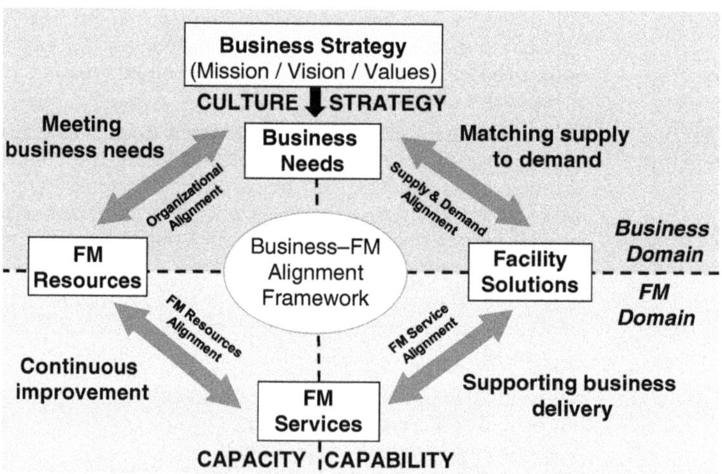

Figure 10.2 Business–facilities management alignment framework

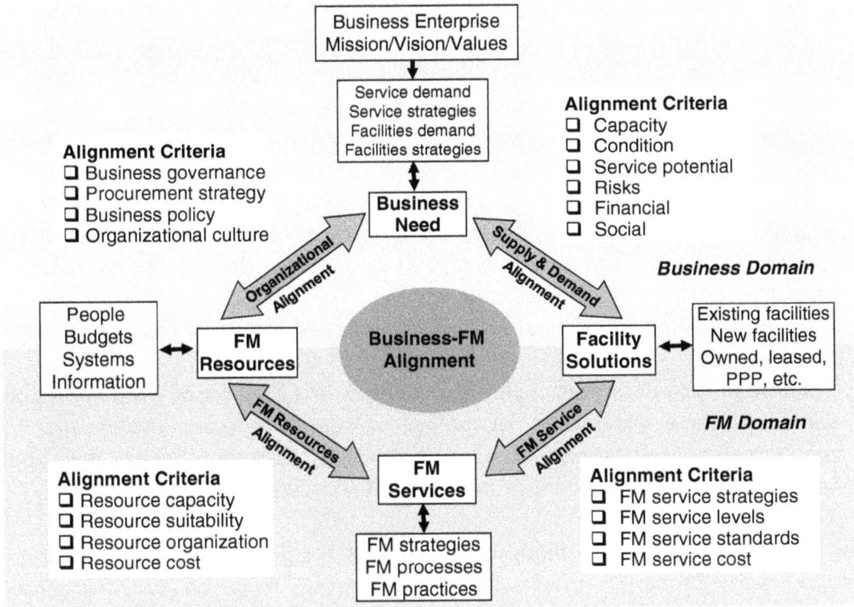

Figure 10.3 Facilities managememt alignment variables and criteria

Taking a vertical delineation, Figure 10.2 also highlights how business organizational culture, as reflected by how FM resources are organized and managed, will impact on its capacity to innovate and continuously improve (left-hand side). Similarly, the quality of facility solutions implemented is a product of appropriate strategy which, in turn, is a function of internal (and external) capability to define, assess and articulate how facilities asset resources and facilities services can best meet planned business objectives.

As shown in Figure 10.3, the proposed business–FM alignment framework has four alignment variables. Each alignment variable is defined by a number of criteria that can be used to achieve or assess the degree of alignment (Table 10.2).

- *Variable (1): Supply and demand alignment – the alignment of business needs with facility solutions*

  The principle that business needs must drive the procurement, ownership and/or leasing of facilities solutions (real estate) is generally accepted. In this context the key variable is that of supply and demand management to ensure that facilities solutions meet business requirements effectively. The alignment of facilities solutions with business requirements is defined as the effective 'fitness' between demand and supply.
- *Variable (2): FM service alignment – the alignment of FM services with facility solutions*

Real estate facilities need to be provided with FM services to support and enable occupancy and use. The suitability of FM services for the type of facilities solutions (e.g. office space, hospitals, factories and schools) is critical in ensuring that the facilities operate effectively and efficiently.

- *Variable (3): FM resources alignment – the alignment of FM resources with FM services*

  Real estate facilities need to be provided with FM services to support and enable occupancy and use. These FM services in turn need to be adequately resourced (people, systems and budgets) in order for them to be effective and meet the needs of the facilities they serve.
- *Variable (4): Organizational alignment – the alignment of FM resources with business needs*

  The FM resources that enable the delivery of FM services need to operate within business parameters of policy, culture, corporate image and the requirements/expectations of the enterprise's customers. This will apply whether the FM services are in-house or outsourced.

The proposed FM–business alignment framework is a product of a combination of comprehensive literature review, expert focus group meetings and case studies reviews. A questionnaire survey was designed to validate the proposed business-FM alignment model and alignment criteria for each arm of the alignment variables. The complexity of the proposed framework dictated that respondents must have relevant management experience and knowledge to be considered as 'experts' in real estate and facilities management. In this respect only senior management personnel were selected as potential respondents. In addition, the scope of geographical coverage of the respondents was international, with respondents from Asia, Australia, North America, Europe and the Middle East. Potential respondents were identified through professional contacts of the researchers and professional institutions.

To date, a sample of 52 respondents from the regions mentioned above, representing a response rate of 49 per cent, have completed the questionnaire. Statistical analyses of the collected responses were performed to check the validity of the proposed alignment framework, alignment variables and criteria. Specifically, the mean values were computed for each alignment criterion and variable. The results of the validation of alignment criteria for each arm of the alignment variables were very promising, with all four arms of the alignment framework having an overall average score of above 4.0 out of a possible maximum of 5.0. The individual scores of each of the criteria within each arm range from a low of 3.37 to 4.73. Efforts are currently under way to extend the survey by increasing the sample size and geographical coverage, and the results will be the subject of further papers which may also feature regional or inter-country comparisons.

There is no doubt that aligning limited resources to achieve corporate objectives is a common concern for corporate resource managers. Our research motivation is to develop a better understanding of how facilities management actions can add

Table 10.2 Alignment criteria – descriptions

Alignment variable	*Alignment criteria*	*Description*
(1) Supply and demand alignment	Capacity	The capacity of the facilities in meeting business requirements in terms of space and other facility attributes that support the delivery of business services or products.
	Condition	The physical condition of the facilities which is appropriate to the functional purpose of the facilities.
	Service potential	This is the potential of the facilities to meet future business requirements and the sustainability of the facilities into the future.
	Risks	These are the risks associated with the facilities which may impact on business productivity and continuity.
	Financial	These are the financial attributes that determine how judgements and decisions will be made in regard to the financial viability of the facilities and include capital value, operating costs and depreciation.
	Social	These are attributes that relate to social responsibility aspects of the facilities, such as heritage and ecological sustainability, to which the enterprise is committed as part of responsible business.
(2) FM service alignment	FM service strategies	These refer to the service strategies developed for the facilities in order to ensure that the facilities are able to perform their intended role in supporting business. They include the service portfolio necessary to achieve the FM service strategy.
	FM service levels	The level of work volume and intensity of each of the FM services that are necessary to meet the requirements of the facility.
	FM service standards	The quality standards of the FM services provided that lead to customer satisfaction.
	FM service costs	The costs of FM services, including procurement, management and transaction costs.
(3) FM resources alignment	Resource capacity	The capacity of the resources to perform the desired functions to the required standard.
	Resource suitability	The suitability of the resources to enable the desired functions to be performed with optimum efficiency and effectiveness.

Table 10.2 (continued)

Alignment variable	*Alignment criteria*	*Description*
	Resource organization	The organization or configuration of the resources to enable the desired functions to be performed with optimum efficiency and effectiveness.
	Resource costs	The costs of the resources, including procurement and management.
(4) Organizational alignment	Business governance	The FM organization structure and the effectiveness of its positioning within the enterprise management and reporting levels.
	Procurement strategy	How FM is sourced as a support function and the extent of compliance with business policy and procurement strategy.
	Business policy	The extent to which the FM functions' operations and services comply with and support business policies.
	Organizational culture	The compatibility of the FM resources (people) with the business environment and culture.

value and contribute towards business success by establishing a structured framework for viewing the alignment 'footprint' of FM within an organization. In this respect, we have begun to shed some light on the following:

- the concept of alignment and how it can assist in improving the appropriate provision of facilities assets and facilities management practice as a strategic and supporting management function;
- the key parameters of facilities assets and facilities services that need to be aligned to business objectives and their relationships;
- the alignment variables between the above parameters and the criteria that further define each of the alignment variables;
- a structured evaluation process or method that can be applied to evaluate the extent of alignment in any given business scenario.

References

Barnhoon, J. G. M. 'Partnerships: A New Management Tool for the Next Decade'. *Facilities*. 1995. Vol. 13, No. 11, pp. 13–18.

Becker, F. and Kelley, T. *Offices at Work: Uncommon Workspace Strategies that Add Value and Improve Performance*. Jossey-Bass. 2004.

Carter, E. 'New Directions in Strategic Asset Management'. 1999. *FM World*. BIFM. Vol. 17, pp. 16–18.

Frost, Y. 'Outsourcing – A Route to Improved Service Delivery?' *International Journal of Facilities*. 1997. Vol. 1, No. 2, pp. 67–76.

Green, A. N. and Jack, A. 'Creating Stakeholder Value by Consistently Aligning the Support Environment with Stakeholder Needs'. *Facilities*. 2004. Vol. 22, No. 13/14, pp. 359–63.

Hastings, N. A. J. *Physical Asset Management*. Springer. 2010. (ISBN: 978-1-84882-750-9).

IBM Institute of Business Value. *A New Way of Working – Insights from Global Leaders*. 2010.

Joroff, M. L. 'Workplace Mind Shifts'. *Journal of Corporate Real Estate*. 2002. Vol. 4, No. 3, pp. 266–74.

Kaplan, R. S. and Norton, D. P. 'The Balance Scorecard: Measures that Drive Performance'. *Harvard Business Review*. Jan–Feb 1992, pp. 71–79.

——'Using the Balance Scorecard as a Strategic Management System'. *Harvard Business Review*. Jan–Feb 1996, pp. 75–85.

——'Having Trouble with Your Strategy'. *Harvard Business Review*. Sept–Oct 2000, pp. 167–76.

——*Alignment*. Harvard Business School Press. 2006.

Krumm, P. J. M. M. and de Vries, J. 'Value Creation through the Management of Corporate Real Estate'. *Journal of Property Investment & Finance*. 2003. Vol. 21, No. 1, pp. 61–72.

Lindholm, A.-L. and Leväinen, K. I. 'A Framework for Identifying and Measuring Value Added by Corporate Real Estate'. *Journal of Corporate Real Estate*. 2006. Vol. 8, No. 1, pp. 38–46.

McBlaine, R. and Moritz, D. 'Transformational Outsourcing: Delivering on the Promise'. *Journal of Corporate Real Estate*. 2002. Vol. 5, No. 1, pp. 357–64.

McDonaugh, J. and Nicola, G. 'Business Strategy and Property Strategy – How Strong is the Linkage?' *Journal of Corporate Real Estate*. 2009. Vol. 11, No. 4, pp. 213–28.

McDougall, G., Kelly, J., Hinks, J., Bititci, U. 'A Review of the Leading Performance Measurement Tools for Assessing Buildings'. *Journal of Facilities Management*. 2002. Vol. 1, No. 2, pp. 142–53.

McGregor, W. and Then, D. S.-S. *Facilities Management and the Business of Space*. Arnold. 1999. Reprinted under Butterworth-Heinemann. 2001. (ISBN: 0-340-71964-8).

Nutt, B. 'Editorial: New Alignments in FM'. *Facilities*. 2004. Vol. 22, No. 13/14, pp. 330–34.

Osgood, R. T. Jr. 'Translating Organisational Strategy into Real Estate Action: The Strategy Alignment Model'. *Journal of Corporate Real Estate*. 2004. Vol. 6, No. 2, pp. 106–17.

——'Key Issues in Enterprise Alignment: Survey Results and Practical Experience with the Fortune 1000'. Flad Architect and CoreNet Global. 2009.

Ratcliffe, J. and Saurin, R. *Towards Tomorrow's Sustainable Workplace: Imagineering a Sustainable Workplace Future*. Johnsons Controls Global WorkPlace Innovation. 2008.

Robertson, K. 'Work Transformation: Integrating People, Space and Technology'. *Facilities*. 2000. Vol. 18, No. 10/11/12, pp. 376–82.

Then, D. S.-S. 'Integrated Resources Management Structure for Facilities Provision and Management'. *Journal of Performance of Constructed Facilities*. February 2003. Vol. 17, No. 1, pp. 34–42.

——'Concepts in Facilities Management'. *Architect's Handbook of Construction Project Management*. Murray, M. and Langford, D. Eds. RIBA Enterprises. 2004, pp. 297–312. (ISBN: 1-85946-123-9).

Then, D. S. S. and Tan, T. H. 'Measuring Operational Building Assets Performance – Concepts and Implementation'. Proceedings of CIBW60–69 Joint Symposium – Measurement and Management of Architectural Value in Performance-based Building. Hong Kong. CIB Publication No. 283. May 2002, pp. 381–96.

——'Aligning Facilities Management Performance to Business Needs – An Exploratory Model Linking FM Performance to Business Performance'. Proceedings of Trondheim CIBW70 International Symposium – Changing Users' Demands in Buildings, Trondheim, Norway. Haugen, T.I., Mourn, A. and Brochner, J, Eds. 12–14 June 2006, pp. 340–49.

Valence, G. 'The FM Industry and Adding Value for Clients'. Brochner, J. and Haugen, T. I. Ed. Proceedings of the 3rd European Research Symposium in Facilities Management. Copenhagen. 12–14 May 2004, pp. 16–22.

Varcoe, B. 'The Performance Measurement of Corporate Real Estate Portfolio Management'. *Journal of Facilities Management*. 2002. Vol. 1, No. 2, pp. 117–30.

11 Summary and conclusion

This book started with the premise that facilities assets, as a component of business resources, are generally undermanaged. To quote Apgar in his latest *Harvard Business Review* article, 'Business real estate is not merely an operating necessity, it's a strategic resource. But it rarely captures senior management's attention' (Apgar 2009, p. 100). As a class of physical assets, they are made up of a complex combination of structures and engineering elements and systems designed to last for many years. Managing them as assets is one of the key elements of facilities management, the other two elements being the management of the workspace within the facilities assets and management of the facilities services required for their upkeep and ongoing operation. By choosing *Facilities Management and the Business of Managing Assets* as the book's title, we have deliberately focused on the asset management aspects of facilities management, with an emphasis highlighting the key essentials of what corporate managers (together with their facilities managers), charged with their provision and ongoing management, should be aware of.

Early emphasis is placed on the crucial relationship with and impact of facilities assets on business performance, with a view to laying the foundation for the approach we have taken in describing facilities asset management. A clear distinction is made between buildings as facilities and other infrastructure assets such as roads and ports, in order to maintain a clear boundary for the scope of discussions.

As long-life durable physical assets, a whole-life-cycle management approach is adopted to develop and describe the principal aspects of facilities asset management. The aim of our book is to provide a level of understanding of facilities asset management for those who have limited knowledge of the subject but who need to bring up their level of knowledge quickly without necessarily having to acquire detailed knowledge of all the processes involved. The fundamental aspects of facilities asset management are outlined in Chapters 4 to 7. Supporting capabilities for the operation of facilities asset management as a business function, such as policies, standards, resources, information and systems, are covered in Chapter 8.

The practice of facilities management is dynamic and will continue to evolve as economic, demographic, business and global issues change and impact on its future development. Therefore facilities asset management and the necessary supporting capabilities have been described in generic terms only. This is in recognition of and belief in the reality that organizations will apply different models of facilities asset management to suit their needs and that these models will have to be flexible and adaptable to change over time. Given that there is ample literature available that describes facilities management processes in detail, our approach has been to provide a high-level management overview of the key elements of facilities asset management. While our aim is to provide a book for corporate managers who are not specialists in the area but may have responsibilities for aspects of facilities management, it also serves as a useful primer for engineers who aspire to be future facilities managers.

Finally, the organization of facilities management as a business function is discussed. In our context, the facilities management organization refers to the business function entity and includes its leadership, management and administrative structure, service delivery model, supporting capabilities and position within the enterprise that it serves. It is the foundation that enables it to define its role and transform itself from a transaction-based function providing cost-sensitive commodity-type services, to a proactive, strategically driven, critical management function that aligns with business strategy and is seen as adding value to the quest for achievement of business objectives.

The dynamic background upon which the role of corporate facilities assets must evolve to accommodate changes in the market place is clearly a constant within business enterprises today. The picture emerging from the research and practice over the last two decades or so is that there is a gradual realization on the part of corporate managers and, at the same time, a growing impatience on the part of real estate/facilities executives, regarding the need for a strategic approach to consideration of issues relating to facilities provision and their ongoing management, as a management process that is integrated with other business support functions. It is hoped that this book has gone some way in providing further impetus towards the pursuit of this goal.

On a broader global scale, the need for facilities and infrastructure assets to be integrated with human society and the natural environment has been glaringly apparent for some time, as demonstrated by the massive surge in infrastructure and facilities development in Asia and other developing parts of the world and their impact on the natural environment. The consumption of non-renewable natural resources, the destruction of historical urban and rural settlements and the displacement of people from their traditional lands have highlighted the need for more responsible planning and management of demand. At the same time, deficiencies in the capacity of governments, business and society generally to adequately maintain and sustain these assets have resulted in many countries with seriously dilapidated infrastructure and facilities assets with few short- or long-term prospects for rejuvenation. This is further compounded when the design and construction of such infrastructure and facilities assets are incompatible with local climatic conditions and the cultural, financial and technical capacity to maintain, operate and use them.

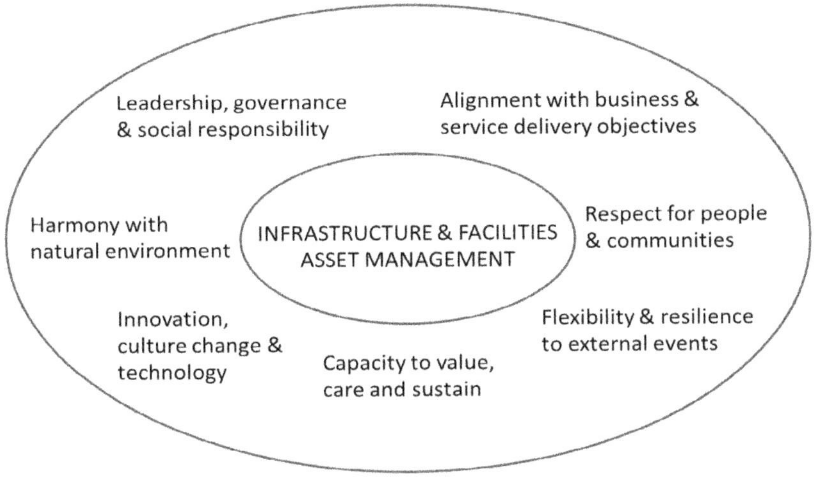

Figure 11.1 Holistic and balanced infrastructure and facilities asset management

Our view is that infrastructure and facilities asset management must evolve towards a more holistic and balanced model in which alignment with business and service delivery objectives is complemented by and integrated with the broader issues of:

- people and society generally;
- harmony with the natural environment;
- flexibility and resilience to external events such as climate change;
- the capacity to value, care for and sustain assets;
- continuous innovation and change towards less asset-intensive solutions through clever use of information, human capital and technology.

The glue that holds such a model together is leadership, governance, accountability and transparency at all levels of government, business and the community. Figure 11.1 illustrates this wider view of infrastructure and facilities asset management.

The development of this model into a successful working framework for implementation is a comprehensive and challenging project that is outside the scope of this book. However, it is hoped that inclusion of this broader, global perspective will at least highlight our recognition that infrastructure and facilities asset management is part of a bigger picture and deserving of further and continuous improvement towards a better world for all. Complacency is not an option.

Reference

Apgar, M., IV. 'What Every Leader Should Know about Real Estate'. *Harvard Business Review*. Nov 2009, pp. 100–07.

Glossary

Accommodation strategy: A strategy that an organization adopts for accommodating its employees and business functions.

Acquisition: In a building context, the purchase, leasing or construction of a facility or building for the purpose of occupancy and use.

Adaptability: In a building context, the ease with which a building can be modified or changed to suit a different purpose.

Alignment: In an FM context, the synchronization of demand and supply (space, services and resources) and integration of organizational culture with business goals for optimum performance.

Alignment variables: In an FM context, the factors that can be assessed to indicate the degree of alignment between alignment parameters.

Assets: In a building and FM context, buildings and their technology components and associated structures that make up the whole functional unit.

Asset life-cycle: In a building context, the stages in the life of a building from planning through design, construction, occupancy and management, renewal and eventual demolition.

Asset management: In a building context, the structured framework of management and technical processes applied to the planning, design, construction, operation, utilization, maintenance, renewal and adaptation of the asset until decommissioning and demolition.

Asset management plan: A plan that sets out how the asset will be managed and maintained through its operation and utilization phase until decommissioning and demolition.

Asset rationalization: In an FM context, the actions implemented for the purpose of adjusting the facilities portfolio to meet new business requirements.

Audits: Critical reviews of performance against agreed criteria.

Benchmark: A standard used as a point of reference.

Benchmarking: Management process used for evaluating business processes and performance against best practice.

Building technology: The electronic information and communications technology that is embedded in the building or used to enhance or enable the operation, maintenance and management of the building and the spaces within.

Business continuity: The ability to continue business processes and operations through a crisis such as a natural disaster, fire or other disruptive events which impact on the business.

Business plans: Documented plans detailing business directions and objectives with specific actions to achieve them within a defined timeframe.

Business resource: Something which helps and supports the business in achieving its goals and objectives, e.g. people, finance, information.

Capital projects plan: Documented plan that details capital building projects to be implemented and the required budgets, proposed start and completion dates and risk management strategies.

Capital replacement: Replacements of major items of building plant and equipment with modern engineering equivalents and usually funded from capital budget allocations.

Climate change: Changes in the earth's climate caused by emissions of greenhouse gases, with the potential to create major impacts on human and natural systems.

Complex: A group of buildings on one or more sites serving a common overall purpose, e.g. hospital, military base. See also 'facility'.

Condition: The physical state of something, e.g. a building, which is normally referenced to an 'as new' baseline.

Corporate real estate: The portfolio of facilities and buildings owned or leased and used by an organization for its own operational purposes.

Demand assessment: Quantitative and qualitative evaluation of the demand for space or services leading to a comparison with the supply available.

Design review: A holistic review of facilities design for the purpose of ensuring that it meets functional requirements, improves construction feasibility and ease of maintenance and achieves the most economical life-cycle costs.

Enterprise: Public sector, not-for profit or private sector organization involved in the provision of goods or services to consumers and/or the community.

Environmental sustainability: The long-term ability of the natural environment to support and provide for human life without permanent degradation of its present and future capabilities.

Facilities asset management: Asset management in the context of facilities. See also 'asset management'.

Facilities assets (or Facility assets): Physical assets that make up the facilities including buildings, plant and equipment.

Facilities audit: An assessment of facilities for the purpose of determining the extent of space utilization, physical condition, functional adequacy and other attributes such as reliability of building services and potential for expansion.

Facilities complexity: As reflected in the design, size and technological sophistication of the buildings comprising the facilities.

Facility: A physical entity constructed and installed to serve a useful purpose. In the context of this book, facility means a single building or a group of buildings serving a common overall purpose, e.g. hospital, military base. See also 'complex'.

Facility alignment plans: Plans for changing the portfolio of exiting facilities to meet new business directions and objectives.

Facility brief: See 'project brief' as applied to facilities.

Maintenance: In a building context, this refers to the work necessary to preserve the integrity of the asset, extend its useful life and maintain functional reliability through scheduled inspections and adjustments, major and minor repairs, restoration and replacements. The works take into account the effects of normal wear and tear and do not materially increase the capital value of the asset.

Maintenance policy: A formal statement which expresses an organization's attitude and commitment towards the maintenance of its physical assets.

Maintenance strategy: The strategy adopted by an organization for the implementation of maintenance in accordance with its stated maintenance policy.

Occupancy costs: The direct and indirect costs associated with occupancy of a building or facility such as maintenance, cleaning, waste management, security, energy, water, telecommunications and building management.

Operating costs: See 'occupancy costs'.

Options analysis: A comparative analysis of the options available with a view to selection of the optimal solution for a particular purpose or issue.

Premises audit: See 'facilities audit'.

Programme management: Management of a group of works directed towards a common purpose, e.g. construction of a number of new facilities within a portfolio spread across a geographical region.

Project brief: Document which expresses the client's requirements, expectations and objectives for a project and is used to control and guide the planning, design and implementation process.

Property: A defined parcel of land with or without any structures built on it.

Real estate facilities: Buildings and other related structures constructed on parcels of land.

Refurbishment: Repairing and restoring to original or 'as new' condition.

Replacement: Exchanging an item of equipment or part of an asset with another that performs the same function. Usually undertaken as a result of failure, deterioration or obsolescence.

Space: In a building and FM context, the area of internal (enclosed) and external (unenclosed) floor surfaces of a building or facility which is used by occupants as part of their normal business or residency.

Space management: The process of managing the spaces within a facility for the purpose of optimizing the efficiency and effectiveness of space utilization.

Space planning: The process of monitoring and assessing trends in demand for space and planning for any anticipated changes in future requirements.

Statutory compliance: Compliance with all relevant laws, regulations, standards and codes.

Supply and demand alignment: A situation where the supply of a particular commodity (e.g. space) matches the demand for that commodity in all respects (e.g. quantity, timing, cost, quality).

Supply chain: The network of businesses that are involved in the sale and delivery of a product from the supplier to the customer.

Testing and commissioning: Testing of plant and equipment to verify their operational readiness and putting them into operational mode for normal operation.

Utilization: In a building and FM context, the occupancy and use of a building or facility.

Bibliography

Books on facilities management

Alexander, K. Ed. *Facilities Management – Theory and Practice.* E & FN Spon. 1998.

Alexander, K., Atkin, B., Brochner, J. and Haugen, T. Eds. *Facilities Management Innovation and Performance.* Spon Press. 2004. (ISBN: 0-415-32146-8).

Atkin, B. and Brooks, A. *Total Facilities Management.* 3rd Edition. Blackwell Publishing. 2009. (ISBN: 978-1-4051-8659-9).

Barrett, P. and Baldry, D. *Facilities Management: Towards Best Practice.* 2nd Edition. Blackwell Science. 2003. (ISBN: 978–0632064458).

Becker, F. *The Total Workplace.* Van Nostrand Reinhold. 1990. (ISBN: 978–0442238117).

Booty, F. *Facilities Management Handbook.* 4th Edition. Butterworth-Heinemann. 2009.

Cooperative Research Centre for Construction Innovation (CRC). *FM as a Business Enabler.* 2007. (ISBN: 978-0-9775282-5-7).

Cotts, D. and Rondeau, E. *The Facility Manager's Guide to Finance and Budgeting.* AMACOM. 2003. (ISBN: 0-8144-0562-2).

Cotts, D., Roper, K. and Payant, R. P. *The Facility Handbook.* 3rd Edition. AMACOM. 2010. (ISBN: 987-0-8144-1380-7).

Finch, E. *Facilities Change Management.* Wiley-Blackwell. 2012. (ISBN: 978-1-1199-6731-6).

Friday, S. *Organization Development for Facility Managers.* AMACOM. 2003. (ISBN: 0-8144-0630-0).

Friday, S. and Cotts D. G. *Quality Facility Management.* John Wiley & Sons. 1995. (ISBN: 0-471-02322-1).

Hinks, J. and Reuvid, J. *Managing Business Support Services.* 2nd Edition. Kogan Page. 2002. (ISBN: 0-7479-3900-9).

McGregor, W. and Then, D. S.-S. *Facilities Management and the Business of Space.* Arnold, 1999; Butterworth-Heinemann. 2001. (ISBN: 0-340-71964-8).

Nutt, B. and McLennan, P. *Facility Management – Risks and Opportunities.* Blackwell Science. 2000. (ISBN: 0-632-05797-1).

RICS. *The Strategic Role of Facilities Management in Business Performance.* 1st Edition (guidance note). August 2009.

www.i-FM.net. *Facilities Management: Realizing the Value of the Investment.* 2011.

——*Facilities Management: New Needs. New Solutions?* 2012.

Books on facilities asset management

CoreNet Global. *Corporate Real Estate 2010 – Integrated Resource and Infrastructure Solutions.* 2004. (ISBN: 978-1-9329-3509-7).

Davis, G. and Venture, F. T. *Performance of Buildings and Serviceability of Facilities.* ASTM. 1990. (ISBN: 0-8031-1292-0).

Hastings, N. A. J. *Physical Asset Management.* Springer. 2010. (ISBN: 979-1-84882-750-9).

Hayes, P. H. and Nunnington, N. *Corporate Real Estate Asset Management: Strategy and implementation.* Elsevier. 2010. (ISBN: 978-0-7282-0575-4).

Lloyd, C. Ed. *Asset Management: Whole-life Management of Physical Assets.* Thomas-Telford. 2010. (ISBN: 978-0-7277-2653-6).

National Research Council. *Core Competencies for Federal Facilities Asset Management through 2020: Transformational Strategies.* Committee on Core Competencies for Federal Facilities Asset Management 2005–20. The National Academies Press. 2008. (ISBN: 978-0-309-11400-4).

Preiser, W. F. E. Ed. *Building Evaluation.* Plenum. 1989. (ISBN: 0-306-43337-0).

Then, D. S.-S. *Real Estate Asset Management – Operational Property Assets and Facilities Support Services as a Business Resource.* Lambert Academic Publishing. 2012. (ISBN: 978-3-8465-2785-6).

Books on space planning and workplace management

Becker, F. and Steele, F. *Workplace by Design: Mapping the High-Performance Workscape.* Jossey-Bass. 1994.

——*Offices at Work: Uncommon Strategies that Add Value and Improve Performance.* Jossey-Bass. 2004.

Best, N., de Valence, G. and Langston, C. Eds. *Workplace Strategies and Facilities Management: Building in Value.* Butterworth-Heinemann. 2003.

McGregor, W. and Then, D. S.-S. *Facilities Management and the Business of Space.* Arnold. 1999; Butterworth-Heinemann. 2003. (ISBN: 0-340-71964-8).

Meel, J., Martens, Y. and Ree, H. J. *Planning Office Spaces – A Practical Guide for Managers and Designers.* Centre for Facilities Management, Technical University of Denmark. 2010.

O'Mara, M. A. *Strategy and Place – Managing Corporate Real Estate and Facilities for Competitive Advantage.* The Free Press. 1999.

Vischer, J. C. *Workspace Strategies – Environment as a Tool for Work.* Chapman & Hall. 1996.

Books on asset maintenance management

Brand, S. *How Buildings Learn – What Happens after They're Built?* Viking Press. 1994. (ISBN 0-670-83515-3).

Boussabaine, A. and Kirkham, R. *Whole Life-cycle Costing: Risk and Risk Responses.* Blackwell Publishing. 2004. (ISBN 1-4051-0786-3).

Chanter, B. and Swallow, P. *Building Maintenance Management.* 2nd Edition. Blackwell Publishing. 2007. (ISBN: 978-1-4051-3506-1).

Ellingham, I. and Fawcett, W. *New Generation Whole-Life Costing.* Taylor & Francis. 2006. (ISBN: 0-415-34658-4).

Kelly, A. *Maintenance Strategy – Business-Centred Maintenance*, Butterworth Heinemann, 1997.

Levitt, J. *The Handbook of Maintenance Management*. Industrial Press. 1997. (ISBN: 0-8311-3075-x).

Wood, B. *Building Care*. Blackwell Science. 2003. (ISBN:0-632-06049-2).

Wordsworth, P. *Lee's Building Maintenance Management*. 4th Edition. Blackwell Science. 2001. (ISBN: 0-632-05362-3).

Index

For Product Safety Concerns and Information please contact our EU representative GPSR@taylorandfrancis.com
Taylor & Francis Verlag GmbH, Kaufingerstraße 24, 80331 München, Germany

www.ingramcontent.com/pod-product-compliance
Ingram Content Group UK Ltd.
Pitfield, Milton Keynes, MK11 3LW, UK
UKHW011019180926
475701UK00004B/841
9780367576547